Communications in Computer and Information Science   371

Pedro Neto António Paulo Moreira (Eds.)

Robotics in Smart Manufacturing

International Workshop, WRSM 2013
Co-located with FAIM 2013
Porto, Portugal, June 26-28, 2013
Proceedings

Springer

Volume Editors

Pedro Neto
University of Coimbra
Department of Mechanical Engineering
Coimbra, Portugal
E-mail: pedro.neto@dem.uc.pt

António Paulo Moreira
University of Porto
Faculty of Engineering
Porto, Portugal
E-mail: amoreira@fe.up.pt

ISSN 1865-0929   e-ISSN 1865-0937
ISBN 978-3-642-39222-1   e-ISBN 978-3-642-39223-8
DOI 10.1007/978-3-642-39223-8
Springer Heidelberg Dordrecht London New York

Library of Congress Control Number: 2013941241

CR Subject Classification (1998): J.7, J.2, I.1.9-10, C.3, F.2, I.5, H.4

Typesetting: Camera-ready by author, data conversion by Scientific Publishing Services, Chennai, India

Printed on acid-free paper

Springer is part of Springer Science+Business Media (www.springer.com)

Preface

This volume contains the papers selected to be presented at the Workshop on Robotics in Smart Manufacturing held during June 26–28, 2013, in Porto, Portugal. It was organized under the umbrella of the 23rd edition of the International Conference on Flexible Automation and Intelligent Manufacturing, FAIM 2013.

This first Workshop on Robotics in Smart Manufacturing brought together people from various sectors (academia, government, and industry), all with a common interest in robotics and manufacturing. Academicians, researchers, students, and industry professionals had the opportunity to discuss and exchange ideas, enlarge networks, and present recent progress and achievements in the emerging areas of application of robots in manufacturing.

The workshop included presentations on the following topics: robotic machining, off-line robot programming, robot calibration, new robotic hardware and software architectures, advanced robot teaching methods, intelligent warehouses, robot co-workers, and application of robots in the textile industry.

Finally, we would like to thank to all the members of the Scientific Committee, reviewers, authors, attendees, FAIM organization staff, and sponsors.

April 2013

Pedro Neto
A. Paulo Moreira

Organization

General Chairs

Pedro Neto   University of Coimbra, Portugal
António Paulo Moreira   University of Porto, Portugal

FAIM 2013 General Chair

Américo Azevedo   University of Porto, Portugal

FAIM 2013 Organization

António Almeida   University of Porto, Portugal

Scientific Committee

Alberto Vergnano   University of Modena and Reggio Emilia, Italy
André G. Conceição   Federal University of Bahia, Brazil
Angel Valera   Technical University of Valencia, Spain
António Paulo Moreira   University of Porto, Portugal
Biao Zhang   US Corporate Research Center ABB INC, USA
Fernando Ribeiro   University of Minho, Portugal
Glauco Caurin   University of São Paulo, Brazil
Gunnar Bolmsjö   University College West, Sweden
Leonardo Honório   Federal University of Juiz de Fora, Brazil
Marcelo Becker   University of São Paulo, Brazil
Mohammad R. Shoaei   Chalmers University of Technology, Sweden
Nuno Lau   University of Aveiro, Portugal
Pedro Fonseca   University of Aveiro, Portugal
Pedro Neto   University of Coimbra, Portugal
Saeid Motavalli   California State University, USA
Sebastian Horbach   Chemnitz University of Technology, Germany

Reviewers

Alberto Vergnano   University of Modena and Reggio Emilia, Italy
André G. Conceição   Federal University of Bahia, Brazil
Andry Pinto   University of Porto, Portugal
Angel Valera   Technical University of Valencia, Spain
António Paulo Moreira   University of Porto, Portugal
Biao Zhang   US Corporate Research Center ABB INC, USA

Dário Pereira	University of Coimbra, Portugal
Diogo Neto	University of Coimbra, Portugal
Eduarda Silva	University of Minho, Portugal
Eduardo Simas	Federal University of Bahia, Brazil
Fernando Ribeiro	University of Minho, Portugal
Germano Lambert-Torres	Itaijuba Federal University, Brazil
Glauco Caurin	University of São Paulo, Brazil
Guilherme Pereira	Federal University of Minas Gerais, Brazil
Gunnar Bolmsjö	University College West, Sweden
Hendrik Hopf	Chemnitz University of Technology, Germany
Hugo Costelha	Polytechnic Institute of Leiria, Portugal
Jorge Ribeiro	University of Porto, Portugal
José Lima	Polytechnic Institute of Bragança, Portugal
Leonardo Honório	Federal University of Juiz de Fora, Brazil
Lingle Wang	Schrodinger, USA
Marcelo Becker	University of São Paulo, Brazil
Marcelo Petry	University of Porto, Portugal
Mohammad R. Shoaei	Chalmers University of Technology, Sweden
Mónica Faria	University of Aveiro, Portugal
Nuno Lau	University of Aveiro, Portugal
Nuno Mendes	University of Coimbra, Portugal
Pedro Costa	University of Porto, Portugal
Pedro Fonseca	University of Aveiro, Portugal
Pedro Neto	University of Coimbra, Portugal
Rúben Oliveira	University of Coimbra, Portugal
Saeid Motavalli	California State University, USA
Sebastian Horbach	Chemnitz University of Technology, Germany
Yuqing Deng	Schrodinger, USA

Sponsors

Faculty of Engineering of the University of Porto
INESCTEC Porto
Portuguese Robotics Society
ROBOPLAN – Robotics Experts
University of Coimbra
University of Coimbra's Mechanical Engineering Research Center, CEMUC®

Table of Contents

An Offline Programming Method
for the Robotic Deburring of Aerospace Components

Francesco Leali, Marcello Pellicciari, Fabio Pini,
Giovanni Berselli, and Alberto Vergnano

"Enzo Ferrari" Engineering Department, University of Modena and Reggio Emilia,
via Vignolese 905/B, 41125 Modena, Italy
{francesco.leali,marcello.pellicciari,fabio.pini,
giovanni.berselli,alberto.vergnano}@unimore.it

Abstract. Deburring of aerospace components is a complex task in case of large single pieces designed and optimized to deliver many mechanical functions. A constant high quality requires accurate 3D surface contouring operations with engineered tool compliance and cutting power. Moreover, aeronautic cast part production is characterized by small lot sizes with high variability of geometries and defects. Despite robots are conceived to provide the necessary flexibility, reconfigurability and efficiency, most robotic workcells are very limited by too long programming and setup times, especially at changeover. The paper reports a design method dealing with the integrated development of process and production system, and analyzes and compares a CAD-based and a digitizer-based offline programming strategy. The deburring of gear transmission housings for aerospace applications serves as a severe test field. The strategies are compared by the involved costs and times, learning easiness, production downtimes and machining accuracy. The results show how the reconfigurability of the system together with the exploitation of offline programming tools improves the robotic deburring process.

Keywords: Offline programming, Industrial Robotics, Integrated Design, CAD-based tools, Digitizers.

1 Introduction

Aerospace cast parts must achieve the design requirements without useless overweight, resulting in complex multifunctional shapes. The manufacturing lots are commonly small and variable but the manufacturing processes must satisfy demanding quality standards [1]. Part deburring is a fundamental process for finishing cast and machined parts to assure the designed performances and to avoid operators hazard [2].

CNC tool machines provide the necessary stiffness and accuracy with low downtimes, high productivity and quality in machining parts. They reduce the damage and rejection of expensive parts and contain the variable labor cost, but involve very high initial investment costs. CNC tool machines also suffer of limited workspace, inflexibility due to limited number of axes and little adaptability capabilities due to the usually rigid tool heads, so that they are not widely used for deburring.

P. Neto and A.P. Moreira (Eds.): WRSM 2013, CCIS 371, pp. 1–13, 2013.
© Springer-Verlag Berlin Heidelberg 2013

A main challenge is the process ongoing adaptation to the high variations of burrs' sizes and profiles geometries [3], so most deburring operations are performed manually with hand held rotary tools. Although a manual method is intrinsically the most flexible, it lacks in safety, time productivity, accuracy and repeatability and must be automated [4]. Moreover, manual operations need very specialized skills to constantly deliver the product quality.

Anthropomorphous robots are the best state of the art compromise between performances and flexibility for automated cast parts deburring [5, 6]. They provide larger work volumes, safety and efficiency at a lower cost than CNC machines. Robots are less stiff and accurate but recent works address accuracy improvements [7]. Also they provide a greater reachability and working capabilities on the complex paths of the deburring tasks [8, 9, 10, 11]. Finally, the anthropomorphous robots lower price and their productivity allow a quick return on investment [12].

Many researches focused on robotic adaptability and changeability, identifying two main strategies to cope with change [13]. Firstly, high system flexibility provides an ongoing adaptation of its behaviors to tasks variations thanks to adjustable controlled equipment. Robotic flexible systems are designed to satisfy a broad range of production requirements and perform a pre-planned, generalized and a-priori flexibility, operating with different products mix, customized sequences and changeable volumes. Robotic flexible systems provide their responsiveness through generally complex hardware and software solutions, and thus not fully exploitable, cost-effective and reliable [4, 14].

The second strategy is proving a reconfigurability customized for product families and specific ranges of production requirements. The reconfigurability paradigm, also defined as the sum of replication and modularization, has addressed computing and robotics industrial research for many years [11]. System is quickly and cost-effectively adapted only when needed, i.e. in event of change of production requirements. Enabler of robotic reconfigurability is the systematic design of modular reconfigurable machines [8], equipment [9] and strategies [10], based on hardware, software and control logics mutually interconnected, e.g. libraries of reconfigurable modules for the integrated description of dynamic behaviours and adaptive control logics [3, 14]. System configuration variations can be produced by using different modules, assembled in different permutations or topologies.

Robotic workcells are basically conceived to be reconfigurable [5] but, at present, robotic manufacturing workcells are often designed to perform complex but repetitive tasks, resulting in dedicated systems with limited changeability. This is a strong limitation when, as in aerospace industry, very complex parts are fed in small lots. In fact at each batch change the workcell has to be reconfigured to execute different work paths with different devices and movement instructions.

Particular attention has to be paid to robot programming, still reported as difficult, time-consuming, and expensive [15, 16, 17]. Point-to-point manual programming stops the production for a long time, depending on the part complexity. It is the more intuitive one but requires specific robot programming skills and long time, since complex profiles often result in hundreds or thousands of points. CAD-based OffLine Programming (OLP) is a state-of-the-art alternative (e.g. [18, 19, 20]), usually faster but must be pursued by a CAD expert rather than a process expert; it is based on perfect geometries of either machining part, robot and the whole cell, so they require an additional real parameters tuning using different information sources. Among the OLP approaches offered on the market, one of the most interesting opportunity is given by touch probes and measuring arms (digitizers), able to generate a robot work path from a physical sample part, instead of a CAD model, not always available. Both CAD-based and digitizer-based OLP are characterized by inherent operative advantages and drawbacks [21].

The present paper discusses an integrated design method for robotic manufacturing systems design and reconfiguration based on OLP tools. Section 2 describes the method and Section 3 exposes its validation through an industrial case study focused on the deburring of cast aerospace housings. The test field consists of contouring very complex profiles. Robot programming is comparatively carried out with CAD-based and digitizer-based approaches. Experimental results are compared in order to reach a significant reduction of the total downtime with a proficient combination of programming and setup methods. Section 4 finally draws conclusions and future works.

2 An Integrated Approach to the Design of Reconfigurable Robotic Workcells

Robot programming has a key role in supporting reconfigurability as a change of the robot tasks due to production variations. To enhance the manufacturing accuracy and reduce the wasted time at changeover, the CAD-based and digitizer-based OLP must follow a structured process.

The method originally applies the fundamentals guidelines firstly proposed by [22] and [23] and synergistically integrates complementary topics to enhance the robotic workcell intelligence: behavioral simulation, OLP, artificial vision, fixed or held compliant deburring tools and grippers [24], [25]. The method focuses on:

1. reconfigurable layout definition, through the design or selection of modular resources according to product and process requirements and specifications;
2. robot OLP, to enable the system reconfigurability and to manage resources interaction;
3. design of modular product/process-dependent hardware and software.

The "Robofacturing" design method workflow is described in Fig.1. The first steps concern the analyses of the industrial problem and the workpiece geometry and material, aiming at clarifying the robotic tasks. This leads to list the end-user requirements and demanded/wished technical specifications, e.g. tolerances and surface quality. General purpose word processors, spreadsheets and 2D object drawing applications represent simple, low cost and flexible instruments for conceptual formalization of the main goals.

Then the manufacturing cycle is conceptually defined. The main robot operations are clarified, logically sequenced and detailed, addressing cutting and finishing tools, manufacturing parameters and checking tool hindrance around workpieces. Since the robot end-effector has an inherent envelope, the area of the workpiece manufacturable by the robot is expressed as the ratio between the surface reachable by the robot and the total surface to be manufactured. Such information are collected in a parametric subprogram module and used in furthers steps of the method.

The next macro-phase is the robot OLP. OLP very reduces the workcell downtimes required by the manual point-to-point robot teaching, but needs a digital description of the part geometries. Two main strategies are here possible: CAD-based and digitizers-based OLP. A CAD-based OLP strategy starts from a 3D model of the workpiece. The model must contain all the information needed to simulate the robot work-cycle: functional geometries for robot picking and part orientation, reference surfaces and planes for the orientation of the robot targets, workpiece and tool reference frames. In a CAD-based OLP two types of software can be used. The first is based on general purpose software platforms, customized with dedicated plug-ins for robotics. They create the work path from the mathematical description of the CAD features. Such "Paths from the Maths" approach is very intuitive and quick but not manufacturing oriented. The second one represents an extension of the typical CAM simulation, where the robot is typically regarded as a 5 + 1 axes tool machine, so that many manufacturing strategies are selectable by the users. Both part-in-hand and tool-in-hand configurations are possible.

A digitizer-based OLP strategy, on the other hand, allows the operators to directly extract the machining features from the real workpiece with a 3D digitizer. Sensorized tips of a multi-DOF arm are manually handled along real part surfaces, recording 3D manufacturing work paths and their reference frames. The work paths can be rapidly loaded in a virtual environment for simulation. The work paths can also be directly converted into the robot code since they can be realized mounting the real tools on the arm and moving it along a real workpieces as in a physical test. Anyway the time saved is compensated by bigger efforts needed during the final workcell tuning since the path is affected by the inherent user inaccuracy.

After robot programming all the designed 3D modular models and subprogram modules are assembled in a virtual workcell layout. Here detail reference frames must be defined and associated to each module to give a reconfigurable architecture to the layout. A workcell virtual prototype is then available to simulate and virtually optimize the process performances and finally generate the robot code.

The number of target points which compose each path has to be determined trading off tolerances, manufacturing cycle times and communication limits of the robot controller. Many control parameters affect such limit and several experimental iterative tests are further needed to find a good compromise between them. Suitable subprograms are concurrently developed for module-robot and/or module-PLC interaction toward I/O connections. The robot manufacturing program is recalled by the workcell program main.

A workcell calibration process is finally required to define the exact position of the reference frames, depending to the real equipment and the robot kinematic and dynamic behaviors. Such step is mandatory for high quality manufacturing. Iterative executions of the robot work-cycle are performed for the final tuning and achievement of all requirements and technical specifications.

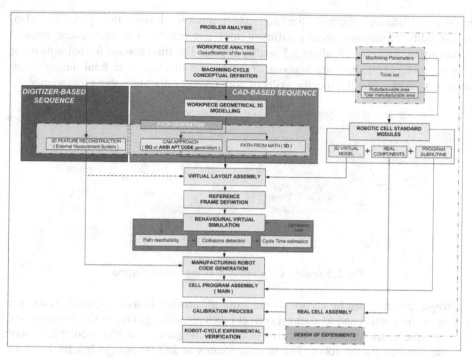

Fig. 1. The "Robofacturing" design method for reconfigurable robotic manufacturing workcells

3   Case Study

OLP plays a key role in the reconfiguration of robotic systems for the aerospace industry, where the small lots production of cast parts demands for frequent changeovers. The present section reports a case study on robotic deburring of aluminum housings for aeronautical gear transmissions. Aeronautic vehicles must be light and performing, and this often leads to more complex large single workpieces designed and optimized to deliver many mechanical functions. They are a good test field for the proposed design method, also due to the close tolerance specifications.

As already shown in Fig.1, the method workflow starts with a preliminary analysis of the robotic deburring problem and workpiece, followed by a first definition of the deburring work-cycle. Then the work paths can be defined with a CAD-based or a digitizer based approach. Such alternatives are evaluated in the following. The workflow goes on with the virtual layout assembly, the workcell calibration and the final tuning addressed in the real robotic workcell.

3.1   Workpiece Description and Robotic Deburring

The shell molded aluminum workpiece is shown in Fig.2. Its envelope is about $220 \times 600 \times 220 mm^3$ and more than 80% of its surface is accessible by the robot. The multi-line profiles to be deburred are positioned on every side of the workpiece, requiring different fixtures for part reorientation during the process. Also, accessibility constraints require different tool geometries. The preexistent manual deburring operations took about 3.5 hours per part. The time needed for point-to-point programming of the housing is about 27 hours, not including the final tuning. Such times very depend on the specific skills of both the operator and robot programmer.

Fig. 2. Sample rotor gearbox housing to be deburred

Irregular burrs must be accurately cut from all profiles to avoid operator risks, not to hinder the gaskets and to assure a perfect lubricant sealing. Due to the dimension of the workpiece, the deburring is performed with the part fixed and a compliant tool spindle handled by the robot. The spindle rotates at 28.000rpm, with a feed rate of 200mm/s with a compliance pressure around 0.2MPa.

The final deburring operation is fed with different raw housings from a product family randomly disposed by an operator on a rotary table. At every work-cycle, the industrial robot ABB IRB 2400/10 handles an artificial vision system above the rotary table to recognize the housing version and its real geometries. Captured information are then matched with the robot database so that grasping coordinates and manufacturing parameters can be loaded into the robot controller and the robot can start its cycle.

The robot moves the housings from the rotary table to the first fixture, then performs an online calibration of the paths and the tool, exploiting all the information captured by the vision system. Reference frames are measured and corrected to automatically compensate misalignments and tool wear occurring in the robotic workcell life-cycle. Finally the robot picks the spindle and the deburring task is run. All the operations are repeated for every placement of the workpiece. Proper carriers collect chips and swarfs while a high power exhaust fan sucks up aluminum dust.

3.2   CAD-Based OLP

The housing geometrical model was realized and optimized through McNeel Rhinoceros and exported as a neutral .iges file. The virtual workcell layout was assembled with mechanical standard modules. Machining parameters and parametric program modules were then coupled with them in order to create an extended database. A behavioral simulation was then easily performed in the virtual environment to check the reachability of the target points, to avoid collisions between the fixed units and the robot and to optimize the cycle time. Compucraft Ltd RobotWorks programming tool, based on Dassault Systèmes SolidWorks CAD, was adopted (Fig.3.).

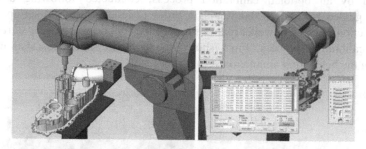

Fig. 3. CAD-based programming environment (Compucraft Ltd RobotWorks)

In the specific case, the spindle orientation was easily defined as orthogonal to the workpiece's surfaces. Since the robot program definition was aided by the software, a deep robot programming language knowledge was not required. A modular program architecture was defined with all deburring subroutines recalled by the main workcell program. The program was finally automatically loaded into the robot controller via Ethernet.

3.3   Digitizer-Based OLP

The 5 DOF digitizer Romer Stinger II was adopted to locate on the workpiece the points necessary to directly reconstruct the robot work paths and their reference planes. Since the digitizing process does not require the real fixtures, the workpiece was referenced only by three very precise column pins. The orientation of each target point depends on the position and orientation of the arm tip during the digitizing phase. So it is useful to directly mount the tools on the digitizers and use them to verify the real tool accessibility and avoid the tool collision against the workpiece. In the deburring of the housing many tools were physically tested: a 90° conical tool, a

Fig. 4. Digitizing of the rotor gear box housing

needle probe, a 6mm spherical probe, an ogive tool and a 6mm cylindrical one. The digitizing operation was performed outside the real workcell, as shown in Fig.4.

The matching between the virtual and the real work paths was afterwards easily guaranteed by an optional calibration process, conducted following a tailored procedure guided by ABB RobotStudio in order to reduce the total set-up time. A short correction must be however realized before the robotic work-cycle starts during the final calibration process. Fig.5 shows the software interface, also used for performing the behavioral simulation.

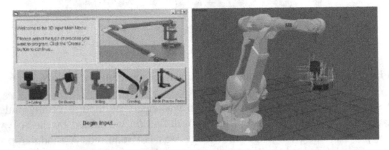

Fig. 5. Calibration software interface for digitizers (ABB RobotStudio)

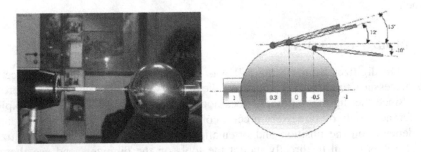

Fig. 6. Touch trigger calibration

The digitizer-based calibration procedure uses a Renishaw MP11 touch trigger. Its real dimensions have to be measured in order to reduce the misalignments between the offline target points acquired by the digitizer and the target positions in the real workcell. First the touch trigger is located in the robot work space and a metal sphere of known diameter is mounted on the robot itself (Fig.6 left). A sequence of points is acquired on the sphere surface, as shown in Fig.6 right. Such procedure is needed in order to measure the real position of the trigger with respect to the robot absolute world reference frame. Then the trigger can be used both for workpiece and tool calibration.

In workpiece calibration (Fig.7), a reference cube of given dimensions is coupled with the workpiece. Then the trigger is mounted on the robot and automatically measures a set of 5 points on every cube surface. Such procedure has to be repeated for all the poses of the workpiece during deburring.

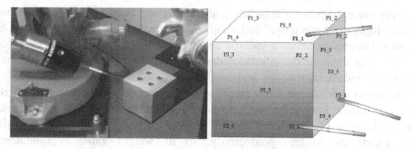

Fig. 7. Workpiece calibration with a reference cube

In a tool-in-hand robot configuration a tool calibration is needed for every tool used during the work-cycle. Calibration is realized adopting a two steps procedure. First a reference cylinder of known dimensions is mounted on the spindle, as shown in Fig.8 at left. A sequence of points is measured on the cylinder (Fig. 8 right), in order to determine the real axis of the spindle and correct any mechanical misalignment at robot flange interface. Then each deburring tool has to be mounted on the moving spindle and measured by the fixed touch trigger.

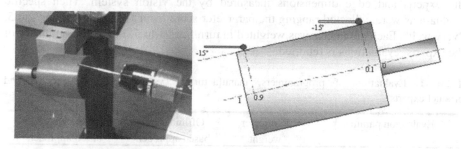

Fig. 8. Spindle and tool calibration with reference cylinder

3.4 Evaluation Test

The two approaches used for the path generation were evaluated by a team of experts. The first evaluations were recorded from interviews, gathered from four programmers and two manufacturing system designers involved in the test, two process and one product experts with CAD skills. The methods are first compared by usability impressions, the involved times and costs, learning easiness and production downtimes. The evaluations consider 10 parameters, as reported in Tab.1, with different evaluation weights.

The evaluation parameter "Cost" refers to the total investment cost of the system composed by both hardware and software. The parameter weight is not maximum because the investment cost is easily compensated by the augmented productivity of the system. "Programming time" considers the time needed for the path generation phase, before entering the virtual assembly environment. The weight of the parameter assumes a medium value because the time does not directly influence the system productivity since the process is mainly realized offline. "Robot language knowledge" and "Manufacturing knowledge" are coupled and deals with the necessity to have preliminary knowledge about a specific robot language and manufacturing process. Such parameters measure the efforts needed for the training of a professional profile expert in both industrial robotics and manufacturing. "CAD knowledge" deals with the preliminary experience in CAD modeling of the operator. The parameter weight is high because a basic knowledge in Computer Aided Design is strongly recommended for the application of the Robofacturing design method. "Production downtime" directly affects the system productivity, so its weight value is fundamental. "Usability" is the parameter considered for the evaluation of the efforts needed for the introduction of the method in an industrial environment. "Final quality (before tuning)" and "Tuning time" have to be considered in order to achieve a qualitative and quantitative understanding of the effectiveness of the two strategies in deburring. The weight of such parameters is clearly very high.

A final campaign was leaded to fine tune the deburring parameters. Tool speed, robot speed, housing position and compliance pressure influences on deburring quality were investigated following design of experiments techniques. Finishing, in terms of absolute quality obtained and process repeatability was finally evaluated by the experts and edge dimensions measured by the vision system. Then specific evaluations were collected ranging the parameter score from 1, meaning "poor", to 5, "very well". Each parameter was weighted in turn from 1 to 5. In Tab.1 a summary of the experts' evaluations is reported.

Table 1. Evaluation of programmers, manufacturing system designers, process and product experts

Evaluation parameter	Parameter weight	Offline CAD based approach	Offline Digitizer based approach
Cost	4	4	1
Programming time	3	4	2
Method learning easiness	3	4	4
Robot language knowledge	5	3	4
Manufacturing knowledge	5	4	3
CAD knowledge	5	1	3
Production downtime	5	4	3
Usability	3	5	5
Final quality (before tuning)	5	4	3
Tuning time	5	4	3
Overall weighted mean	-	**3.6**	**3.1**

4   Conclusions

Robot OLP has the great advantage of reducing the overall production downtimes. In fact the only phases booking the robot are the workcell calibration and the final tuning. OLP has been integrated in an extended design method for reconfigurable manufacturing workcells.

Two different OLP approaches were compared: a full virtual one based on CAD tools and a second one which starts from data acquired by a measuring arm. Paths and program generation with a CAD based approach are very fast and it is easy to verify the cycle times and target point reachability for different working strategies. The virtual model includes also mechanical devices and fixtures so it is possible to detect collisions and optimize the whole robot work-cycle. The main drawbacks are that the operator must be quite skilled in the use of specialized software and the calibration is difficult since the program considers perfect geometries for robots, tools, fixtures and workpieces without any geometry variability. Moreover, user training is heavy and time consuming. Since it can be used for different robot brands, the possible lack of integration between the virtual controllers, based on Realistic Robot Simulation (RRS) algorithms, and the real robot controllers could affect the robotic machining accuracy. The total required time was about 3hours for a simplified workcell CAD modeling and paths generation and program post-processing, not including the final workcell tuning.

On the other hand the digitizing approach doesn't need a CAD model, advances a preliminary calibration on real geometries and a direct verification on the work path reachability. The calibration itself is semi-automated and many features are available together with the measuring arm for the path acquisition. Such approach does not require advanced programming skills and the program generation is quite fast. On the other hand the tools can be used only for vendor specific robots and the procedure is longer than the CAD-based one. The user training is simplified because of the intuitive interaction with the real workpiece and tools. A good skill in operating both in a real and virtual environment is mandatory. The required times were: 1hour for test setup and calibration with the measuring arm, 3hours for offline points acquisition and robot programming and for probe, spindle and workpiece calibration. The total required time was 6.5hours, not including the final workcell tuning.

In both the approaches followed for the work path generation, the deburred edges present a sufficient uniformity and the size of the contouring profile is within the given tolerances. Anyway the target quality is achieved thanks to a specific tuning realized in the real workcell at every changeover.

Acknowledgments. The authors want to acknowledge Luciano Passoni, Davide Passoni and Lino Ferrari, with SIR S.p.A. (Modena, Italy), for their technical and managerial contribution to the project, and AVIO S.p.A. (Turin, Italy) for supporting the experimental tests.

References

1. Davis, J.R.: Cast Irons. ASM International (1996)
2. Aurich, J.C., Dornfeld, D., Arrazola, P.J., Franke, V., Leitz, L., Min, S.: Burrs—Analysis, control and removal. CIRP Annals - Manufacturing Technology 58, 519–542 (2009)
3. Ahmad, M.M., Sullivan, W.G.: Flexible automation and intelligent manufacturing. Robotics and Computer-Integrated Manufacturing 18(3-4), 169–170 (2001)

4. Tolinski, M.: Deburring Processes and Challenges. Manufacturing Engineering 137(4), 83–97 (2006)
5. Chen, S.-C., Tung, P.-C.: Trajectory planning for automated robotic deburring on an unknown contour. International Journal of Machine Tools and Manufacture 40(7), 957–978 (2000)
6. Sugita, S., Itaya, T., Takeuchi, Y.: Development of robot teaching support devices to automate deburring and finishing works in casting. International Journal of Advanced Manufacturing Technology 23, 183–189 (2004)
7. COMET Project – Plug-and-Produce COmponents and METhods for Adaptive Control of Industrial Robots Enabling Cost Effective, High Precision Manufacturing in Factories of the Future. In: European 7th Framework Programme, reference number 258769, http://www.cometproject.eu
8. Meziane, F., Madera, S., Kobbacy, K., Proudlove, N.: Intelligent systems in manufacturing: current developments and future prospects. Integrated Manufacturing Systems 11(4), 218–238 (2000)
9. Kramer, B.M., Shim, S.S.: Development of a system for robotic deburring. Robotics & Computer-Integrated Manufacturing 7(3-4), 291–295 (1990)
10. Schimmels, J.M.: Multidirectional compliance and constraint for improved robotic deburring. Part 1: improved positioning. Robotics and Computer-Integrated Manufacturing 17(4), 277–286 (2001)
11. Schimmels, J.M.: Multidirectional compliance and constraint for improved robotic deburring. Part 2: improved bracing. Robotics and Computer-Integrated Manufacturing 17(4), 287–294 (2001)
12. Whitney, D.E.: Research issues in manufacturing flexibility - an invited review paper for ICRA 2000 symposium on flexibility. In: The IEEE International Conference on Robotics and Automation, pp. 383–388. IEEE Press, S. Francisco (2000)
13. Ollero, A., Boverie, S., Goodall, R., Sasiadek, J., Erbe, H., Zuehlke, D.: Mechatronics, robotics and components for automation and control. IFAC Milestone Report, Annual Reviews in Control 30(1), 41–54 (2006)
14. Andrisano, A.O., Leali, F., Pellicciari, M., Vergnano, A.: Engineering Method for Adaptive Manufacturing Systems Design. International Journal on Interactive Design and Manufacturing 3, 81–91 (2009)
15. Dietz, T., Schneider, U., Barho, M., Oberer-Treitz, S., Drust, M., Hollmann, R., Hägele, M.: Programming System for Efficient Use of Industrial Robots for Deburring in SME Environments. In: 7th German Conference on Robotics, ROBOTIK 2012, pp. 428–433. VDE-Verlag, Munich (2012)
16. Wu, B.: Manufacturing strategy analysis and system design – the complete cycle within a Computer Aided Design environment. IEEE Transactions on Robotics and Automation 16(3), 247–258 (2000)
17. Mitsi, S., Bouzakis, K.-D., Mansour, G., Sagris, D., Maliaris, G.: Off-line programming of an industrial robot for manufacturing. International Journal of Advanced Manufacturing Technology 26, 262–267 (2005)
18. Chen, H., Fuhlbrigge, T., Li, X.: A review of CAD-based robot path planning for spray painting. Industrial Robot 36(1), 45–50 (2009)
19. Kim, J.Y.: CAD-based automated robot programming in adhesive spray systems for shoe outsoles and uppers. Journal of Robotic Systems 21(11), 625–634 (2004)
20. Neto, P., Mendes, N., Araújo, R., Pires, J.N., Moreira, A.P.: High-level robot programming based on CAD: dealing with unpredictable environments. Industrial Robot 39(3), 294–303 (2012)

21. Pan, Z., Polden, J., Larkin, N., Van Duin, S., Norrish, J.: Recent Progress on Programming Methods for Industrial Robots. Robotics and Computer-Integrated Manufacturing 28, 87–94 (2012)
22. Pahl, G., Beitz, W.: Engineering design: a systematic approach, 2nd edn. Springer (1996)
23. Boothroyd, G., Dewhurst, P., Knight, W.A.: Product design for manufacture and assembly, 2nd edn. CRC Press Publisher (2001)
24. Pellicciari, M., Leali, F., Andrisano, A.O., Pini, F.: Enhancing Changeability of Automotive Hybrid Reconfigurable Systems in Digital Environments. International Journal on Interactive Design and Manufacturing 6, 251–263 (2012)
25. Andrisano, A.O., Leali, F., Pellicciari, M., Pini, F., Vergnano, A.: Hybrid Reconfigurable System Design and Optimization through Virtual Prototyping and Digital Manufacturing Tools. International Journal on Interactive Design and Manufacturing 6, 17–27 (2012)

Experimental Investigation of Sources
of Error in Robot Machining

Ulrich Schneider[1], Matteo Ansaloni[2], Manuel Drust[1],
Francesco Leali[2], and Alexander Verl[1]

[1] Department Robot and Assistive Systems, Fraunhofer Institute
for Manufacturing Engineering and Automation IPA, Nobelstrasse 12,
D-70569 Stuttgart, Germany
{Ulrich.Schneider,Manuel.Drust,Alexander.Verl}@ipa.fraunhofer.de
[2] Department of Engineering "Enzo Ferrari", University of Modena and Reggio Emilia
Strada Vignolese 905 – 41125 Modena, Italy
{Matteo.Ansaloni,Francesco.Leali}@unimore.it

Abstract. This document is divided into two parts. First a survey is given presenting sources of error in robot machining and outlining their dependencies. Environment dependent, robot dependent and process dependent errors are addressed. The second part analyses the errors according to their source, magnitude and frequency spectrum. Experiments under different conditions represent a typical set of industrial applications and allow a qualified evaluation. This analysis enables the qualified choice of suitable compensation mechanisms in order to reduce the errors in robot machining and to increase machining accuracy.

Keywords: robotic machining, robot dynamics, robot precision, robot compensation.

1 Introduction

Industrial robots (IR) are traditionally used for handling applications. According to the International Federation of Robotics 78% of all industrial robots were used for handling and welding [1]. As the demands for more flexibility and lower costs are rising in industry new concepts have to be developed to satisfy the requirements of modern production. As industrial robots offer lower costs than a conventional tooling machine, an exceptional flexibility and a big working area, more and more industrial robots are used for machining operations. However, so far industrial robots cannot compete with machine tools in the field of high precision. Due to a large set of error sources, industrial robots cannot address the same variety of applications like conventional machine tools. Machine tools are optimized for the machining process by providing high stiffness. Yet industrial robots are originally conceived to do handling operations and provide a large work space.

P. Neto and A.P. Moreira (Eds.): WRSM 2013, CCIS 371, pp. 14–26, 2013.

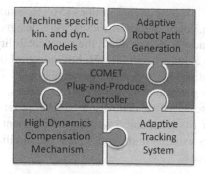

Fig. 1. Concept of accurate machining with industrial robot in COMET [2]

The traditional six rotational axes enable a big flexibility because of the vast set of positions and orientations which can be targeted. As this serial design of industrial robots enables several advantages on the same time the serial chain of joints is the major disadvantage of robots when trying to compete with machine tools. The errors of all joints sum up to the tool centre point (TCP) and reduce the precision of the robot. As soon as higher precision are needed robots are replaced by traditional machine tools. Therefore a much larger set of applications could be addressed by robots if the accuracy could be increased. One possible approach to increase robot precision is the precise description of the sources of error and their usage for compensations. This topic is dealt within the EU/FP7-project COMET [2].

The sources of error in robot machining are investigated and compensation mechanisms are set up in order to increase to accuracy in the machining process. The combination of different compensation approaches aim at a guaranteed accuracy of 50 µm. The target system is demonstrated in Fig. 1. Four steps are taken towards accurate robot machining: Model-based compensation is combined with a holistic programming approach. The additional tracking of the robot's TCP allows to feed the reference back to the robot controller and to an external high dynamic compensation mechanism compensating for the frequencies exceeding the bandwidth of the robot [3], [4]. This paper is organized as follows. Section 1 describes the relevance of robot machining and the impact of the achieved accuracy. A survey on sources of error is given in Section 2. Section 3 gives a detailed analysis of the characteristics of influences and provides dependencies. After presenting an overview over compensation strategies and evaluating the effects in Section 4 the paper finishes with conclusions and an outlook in Section 5.

2   Survey of Sources of Error in Robot Machining

Among the different performances related to the robot itself, precision is often used to describe its capabilities, and is further divided in: repeatability, accuracy and resolution. Repeatability and accuracy estimate the closeness between a set of attained positions and orientations of the TCP, when repeating the robot motions into the same commanded pose and their nominal values [5]. Resolution encompasses also programming resolution. Since industrial robots were designed to execute repeatable operations, their accuracy is lower than their repeatability. A typical industrial manipulator accuracy is about ±1 mm [6], [7], but values of 0.3 mm could be reached with accurate compensation [8]. Repeatability ranges in 0.1 - 0.03 mm [9]. Errors are responsible for the degradation of these performances. In order to obtain a clarification of the sources, a first distinction can be carried out among sources of error in the robot itself (its mechanical structure, basement and control system) or robot dependent, sources external to the robot (cell and auxiliary devices) and process (or task) dependent sources.

2.1   Environment Dependent Errors

The real accuracy of a robot depends strongly on the full chain of components between the tool on the TCP and the floor. Starting from the environment the structure of the building has an impact on the behavior of the robot. The presence of a basement changes the transmission from the environment on the robot. Especially, when measuring in the range of μm those effects cannot be neglected. In Fig. 2 a typical situation for a production facility environment is considered, disturbances arising from a pallet truck and passing people are applied. The signals are the relative movement between the Keyence sensor, with an accuracy of 1 μm, and the robot which are both attached on the 14 tons machine bed. An FFT of the signals reveals the main resonances to be similar to the resonances of the robot. It can be concluded that the measured signal is a real movement of the robot due to disturbances from the environment.

The chain of transmission of disturbances continues with the material of the floor and the fixture of the robot to the floor. Due to the big lever from the base to the tool small deformations in the base lead to big deviations on the tool. Moreover, an influence which must not be neglected is temperature. Different materials with different coefficients are used within an industrial robot which leads to a deformation which is hard to predict [11]. Also the tool holder and spindle support compliance must be taken into account. In general their contributions to the compliance of the system cannot be neglected. Further, cell calibration is another important issue which directly effects the final quality achieved. In robotic machining, the cell environment replaces the machine tool basement and the fixturing feature of the latter should be replaced with dedicated devices. In modern robotic cells, offline robot programming methods are used when the robot tasks are complex, as in the case of robot path required for machining. The current practice of creating the robot path with the aid of CAD/CAM software [2] requires a close matching between the CAD representation of the workcell and its real environment. Current approaches are based on CAD knowledge of the cell, devices (e.g. tool holder) and robot, which provide extreme flexibility but impose to adopt further calibration strategies to fulfill process accuracy requirements. Following the common approach of cell calibration position and orientation of cell components are computed using vision-based automated algorithms [12], [13].

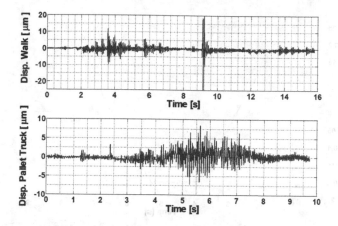

Fig. 2. Influence of disturbances measured on a 14 tons machinebed

2.2   Robot Dependent Errors

Within the mechanical robot structure two categories of errors can be distinguished: Geometrical errors and non-geometrical errors [8]. The former encompasses all the deviation due to imperfect geometries, mating or assembly errors, and these errors exist whether the robot is moving or not. The latter include all the sources related to the dynamical behavior of the robot. In addition, unlike the former, they are time-varying and change in magnitude during manipulator operations. The main effect of both of the sources is causing discrepancies between the real robot and its kinetostatic and dynamic model from which its characteristics are derived [14] and on which control is based [15].

a) Geometrical errors: Geometrical errors, which are generally compensated by calibration, arise from manufacturing or machining tolerances of robot components. Tolerances introduce variations in link geometry, as well as some variation in the orientation of the joints after link assembly and nonlinearities in the gears. Then, these errors will propagate to cause inaccuracy in the pose of the TCP. Links tolerances are not the unique source of geometrical errors. Joint errors in the axes are produced during the assembly of the various joint components due to clearance in motor and geared transmission mechanisms, backlash and bearing run-out errors. Backlash effects are a function of the geometrical looseness of the gears produced when they are mated together. These errors can make a significant contribution, even larger than that due to geometric tolerances, to robot positioning accuracy [16], [17]. Yet as in robot machining initially a cell calibration and a referencing procedure of the real position of the work piece are performed, only nonlinearities of the gears are considered in this paper. As the robot locally shows a rather good accuracy the impact of most geometrical errors (except nonlinearities of gears) can be reduced to errors in tool calibration (position of the tool on the TCP of the robot) and nonlinearities and measurement errors of the applied sensors. Yet these errors may vary depending on the individual circumstances as disturbances such as dust, conservation liquid and burrs of the workpiece may introduce additional errors.

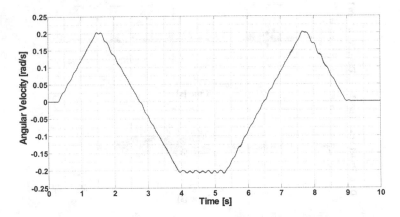

Fig. 3. Influence of compliance and backlash of axis 1

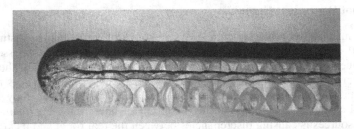

Fig. 4. Impact of gear backlash when machining in aluminium

b) Non-geometrical errors: Non-geometric errors also occur in a local environment and therefore cannot be compensated for by cell calibration. They arise from structural deformations of load-transmitting components, links and energy-transforming devices, wear and nonlinear effects such as nonlinear stiffness, stick-slip motion and hysteresis in servodrives [18], [19]. The compliance errors are due to the compliance of the links and joints under inertial and external load. In particular, joint compliance results from the torsional stiffness of the gearbox and the output drive shaft actuating the joint. Besides, the masses of the links cause an additional torque on the gears due to gravity effects. Especially during machining, forces add on the load of the gears and cause additional deflection. Link and joint compliance, causing the deflection of the links and finally the TCP, contribute up to 8-10% of the position and orientation errors of the TCP [8].

In addition, joint, and to a less extent links, compliance cause vibrations of the robot structure during its movements. Especially, when the industrial robot is driven with high speed, the industrial robot has large vibrations caused by the speed reduction mechanism [20]. Moreover, when the load on the TCP changes rapidly, or robot is undergoing fast movement, the resonant phenomenon will appear. Compliance and backlash are the two most effective influences of a robot's gears and drives. The natural damping of such systems is very low and yields to a slow decay characteristic of torsional oscillations [19], [21], [22]. In addition backlash yields too high torque impulses which can excite torsional vibrations. In Fig. 3 these effects of first axis measured on the TCP of a KR125 are demonstrated. Further measurements on the stiffness of joint 1 allowed the identification of its compliance as well as the identification of the backlash value. The results are: 0.9° for the backlash and $3.6 \cdot 10^6$ Nm/rad for the compliance. Machining experiments in aluminum show the great impact of backlash (see Fig. 4). The exemplary compliance of axis 1 is measured. The results are presented in Fig. 5. Assuming a lever of 1.5 m a realistic load on the TCP of 300 N caused by a machining process would result in a torque of 450 Nm and a deflection of 0.2 mm. In robotic machining process, the force induced deflection of the robot structure is the single most dominant source of error. Even though all components of an industrial robot contain intrinsic compliance, the major compliance can be assigned to the gears. Other important sources of error inside the mechanical structure are wear of the parts, the internal heat sources such as motors and bearings. Wear of the parts is strictly related to friction, in particular stiction, which in turn depends on temperature, joint applied torque and rotating speed [23].

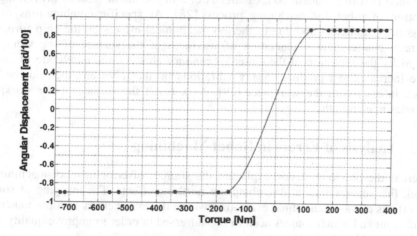

Fig. 5. Compliance measurements of axis 1 of a KR125

c) System Errors: Errors in this category include those caused by improper calibration, sensor measurement errors, control implementation errors and numerical round-off errors in the computer used for control. Sensor error is due to the joint angle sensor resolution and mounting. Due to the biggest lever axis 1 has the biggest impact on the TCP. When positioned in machining configuration minimal movements of 2 μm could be identified on the TCP. Control and algorithmic errors are related to the geometrical model implemented in the controller. Especially for model-based controls precise and accurate models of the nonlinearities are required [24]. Furthermore, also the controller sampling time contributes to these errors especially in a real-time context [25].

2.3   Process Dependent Errors

In machining applications the most important position source error is the machining force induced error. The machining force in an aluminum-milling process is hundreds of Newton, consequently the force induced error reaches values up to 1 mm [26] (compare Fig. 7 and Fig. 8). The structure of the robot transmits this force to the workpiece according to its mechanical characteristics. The values of the machining forces depend on the process parameters: spindle speed, axial depth-of-cut, radial depth-of-cut and chip load. They result in a specific value for the material removal rate value (MRR). In traditional machining application, feed is kept constant in spite of the variation of depth of cut and width of cut [27]. This will introduce a dramatic change of MRR, which would result in heavy changes in the machining force. The lubrication system is another important factor, especially for the final quality of the workpiece. The lubricating oil reduces the contact friction coefficient between the workpiece and the cutter, moreover this contributes to avoid the first type of chatter. The effects are measurable on the final quality surface of the machined part (e.g. roughness). Chatter is one of the major reasons preventing the adoption of robot for machining process [28]. At specific combinations of the foregoing parameters and due to thermo-mechanical effects on the chip formation (primary chatter) and regeneration of waviness of the surface of the workpiece caused by the vibration of the cutter (secondary chatter), the amplitude of cutting force increases and produces heavy vibrations on the robot and then on the TCP which interact with the workpiece [29]. As a result the surface of the workpiece becomes non-smooth.

3   Analysis of Errors in Robot Machining

Whereas the previous section explained the sources of error in robot machining in detail, this section aims at describing the resulting effects. The mapping of sources and effects allows then a final evaluation where the major errors in robot machining result from and which sources need to be addressed in order to improve quality when machining with industrial robots.

3.1   Experimental Setup

A KR125 from KUKA is used for the experiments (see Fig. 6). It is driven by a Beckhoff TwinCAT CNC and therefore optimized for the machining process. A Chopper 3300 spindle from Alfred Jaeger is used together with a 8 mm end mill tool with four teeth from Hoffmann Group. A Leica Absolute Tracker AT901 is used to measure the robot behavior and to determine parameters of the error sources. The tracker can perform three dimensional measurements at 1 kHz, with an error of ErrLT < 20 µm for the chosen area. A one-dimensional Keyence LK-G87 laser triangulation sensor is used in order to capture the influences of the surroundings on the robot. Robot and spindle are mounted on a 14 tons machine bed in order to decouple the cell from the surroundings. The lab is on first floor over the basement.

Fig. 6. Experimental setup with machine bed, KR125 robot, Chopper 3300 spindle, Keyence LK-G87 sensor and Lasertracker

3.2   Robot in Machining Operation

The robot in a machining operation is a complex system. The characteristic vibrations of the robot are combined with the oscillations due to the machining process. A machining example in ST-37 steel is chosen in order to demonstrate the typical effects in robot machining. The spindle speed is set to 10000 rpm and the feed is defined as 1000 mm/min. Using a tool with four teeth, the process parameters allow to evaluate the fundamental tooth passing (or first harmonic) frequency which value is f = 666.7 Hz. Machining is performed in full width cut. As the robot shows different properties when moving in different directions two experiments were performed:

- Machining a straight line following the y-axis
- Machining a straight line following the z-axis

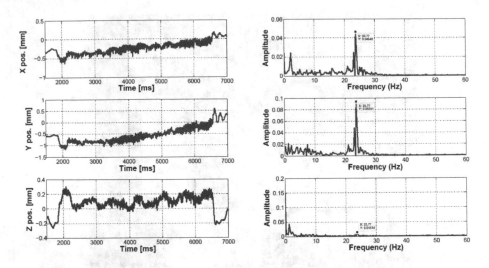

Fig. 7. Position and FFT when machining in y-direction

First of all the deflection of the robot when entering the material should be pointed out. In full width cut process forces are most present in feed direction and orthogonal to feed [30]. Due to the limited stiffness and these process forces the robot is deflected from its targeted path (Fig. 7 and Fig. 8). As the robot in the used configuration is much more compliant in z-direction than in y-direction the deflection orthogonal to path when machining in y direction is bigger. Secondly, the frequency analysis of the signal shows interesting results. As the attachment of the spindle is

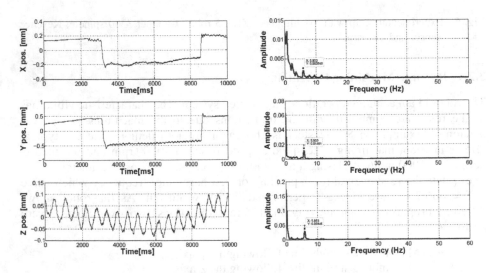

Fig. 8. Position and FFT when machining in z-direction

considered to be stiffer than the robot all lower frequencies can be assigned to the robot. It is obvious that the dominant frequency can be found at 5.93 Hz and 23.77 Hz. As the two machining scenarios cover the most compliant and the stiffest configuration of the robot it can be concluded that the bandwidth of the robot varies between these two values depending on its configuration. Finally also the nonlinearities of the gears are clearly visible with an amplitude of ±0.1 mm. They do no change with the speed of the robot but they show up as a low frequency in the FFT in the experiment. However, they do not limit the bandwidth of the robot but influence only the accuracy of the robot.

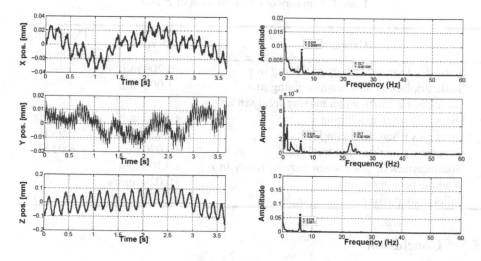

Fig. 9. Position and FFT when moving in z-direction

3.3   Robot in Free Space Motion

In contrast to a robot in machining a robot in free space movement is not excited by external disturbances. When moving the TCP in z-direction the impact of compliance and backlash of all axes result in the characteristic eigenfrequencies already experienced in machining (compare Section 3). Fig. 9 shows position and frequency properties of the free space motion. It should be noted that not only frequencies but also amplitudes of the oscillations in machining and in free space motion are comparable. As expected the nonlinearities of the gears appear like in the machining experiment.

4   Summary

According to the previous sections it can be concluded that the dominant frequencies in robot machining only depend on the mechanical properties of the robot. The effects can be traced back to the compliance and the backlash of the gears determining the frequency of position disturbances in the TCP. The results of all measurements describing the effects on the TCP are summarized in Table I. As expected, changing the configuration of the robot leads to different proprieties in terms of compliance and natural frequencies. This can be easily recognized in the final surface finishing (compare Fig. 4).

As in machining the exciting frequencies are always higher than the eigenfrequencies of the robot (compare section 3.2) the robot is very likely to oscillate with its eigenfrequencies. This mechanical constraint can only be influenced by mechanical modifications or overcome by external actuation like it is presented in [3], [4].

Table 1. Summary of effects in robot machining

Experiment	Deviation	Dominant frequency
Static displacement when machining in y	0.200 mm	-
Static displacement when machining in z	1.000 mm	-
Static displacement when moving freely in z	-	-
Dynamics when machining y	±0.250 mm	23.77 Hz
Dynamics when machining z	±0.050 mm	5.93 Hz
Dynamics when moving freely in z	±0.070 mm	6.02 Hz
Nonlinearities of gears when moving freely in z	±0.100 mm	-
Walking person passing	±0.020 mm	-
Pallet truck passing	±0.007 mm	-

5   Conclusion

The presented paper analyses the relevant sources of error when machining with industrial robots. The full mechanical chain from the environment to the flange including the robot controller was considered. The most important sources were identified and quantified. Experiments in machining and experiments in free space motion show that compliance and backlash are the most dominant sources. However when trying to achieve an accuracy of < 100 μm also the disturbances from the environment and errors from cell calibration need to be taken into account. Position and frequency analysis demonstrate the dependency on the robot configuration and identify the stiffest configuration of the robot. Based on the analysis a compensation of compliance and backlash can be identified as being most effective. Calibration of the robot kinematics and the calibration of the workcell can improve positioning accuracy and results also in better precision in machining. Proper decoupling of the cell components from the environment and from each other can reduce process disturbances further. The intrinsic oscillation of a serial robotic system can only be eliminated by external devices.

Acknowledgments. The research leading to these results has received funding from the European Union's seventh framework program (FP7/2007-2013) under grant agreement #258769 COMET.

References

1. International Federation of Robotics: World Robotics 2007. Statistical Yearbook (2008)
2. COMET, 2011, EU/FP7-project: Plug-and-produce COmponents and METhods for adaptive control of industrial robots enabling cost effective, high precision manufacturing in factories of the future, http://www.cometproject.eu (accessed January 2013)
3. Puzik, A.: Genauigkeitssteigerung bei der spanenden Bearbeitung mit Industrierobotern durch Fehlerkompensation mit 3D-Ausgleichsaktorik. Dissertation, University of Stuttgart, Fraunhofer IPA (2011)
4. Puzik, A., Pott, A., Meyer, C., Verl, A.: Industrial robots for machining processes in combination with an additional actuation mechanism for error compensation. In: Proceedings of 7th Int. Conf. on Manufacturing Research (ICMR), Warwick (2009)
5. Standard ISO 9283: Manipulating industrial robots – Performance criteria and related test methods (1998)
6. Siciliano, B., Kathib, O.: Handbook of Robotics. Springer, New York (2008)
7. Shiakolas, P.S., Conrad, K.L., Yih, T.C.: On the accuracy, repeatability, and degree of influence of kinematics parameters for industrial robots. International Journal of Modelling and Simulation 22, 245–254 (2002)
8. Mustafa, S.K., Pey, Y.T., Yang, G., Chen, I.: A Geometrical Approach for Online Error Compensation of Industrial Manipulator. In: Proceedings of IEEE/ASME International Conference on Advanced Intelligent Mechatronics, Montreal, Canada, July 6-9, pp. 738–743 (2010)
9. Breth, J.F., Vasselin, E., Lefebvre, D., Dakyo, B.: Determination of the Repeatability of a Kuka Robot Using the Stochastic Ellipsoid Approach. In: Proceedings of IEEE International Conference on Robotics and Automation Barcelona, Spain, pp. 4339–4344 (2005)
10. Elatta, A.Y., Gen, L.P., Zhi, F.L., Daoyuan, Y., Fei, L.: An Overview of Robot Calibration. Information Technology Journal 3, 74–78 (2004)
11. Heisel, U., Richter, F., Wurst, K.-H.: Thermal behavior of industrial robots and possibilities for errors compensation. CIRP Annals - Manufacturing Technology 46, 283–286 (1997)
12. Pan, Z., Polden, J., Larkin, N., Van Duin, S., Norrish, J.: Recent progress on programming methods for industrial robots. Robotics and Computer-Integrated Manufacturing 28, 87–94 (2012)
13. Tarn, T.J., Song, M., Xi, M., Ghosh, B.J.: Multi-Sensor Fusion Scheme for Calibration-Free Stereo Vision in a Manufacturing Workcell. In: Proceedings of IEEE International Conference on Multisensor Fusion and Integration for Intelligent Systems, pp. 416–423 (1996)
14. Legnani, G., Tosi, D., Fassi, I., Giberti, I., Cinquemani, S.: The "point of isotropy" and other properties of serial and parallel manipulators. Mechanism and Machine Theory 45, 1407–1423 (2010)
15. Dietz, T., Schneider, U., Barho, M., Oberer-Treitz, S., Drust, M., Hollmann, R., Hägele, M.: Programming System for Efficient Use of Robots for Deburring in SME Environments. In: Proceedings of 7th German Conference on Robotics (2012)
16. Oh, Y.T.: Influence of the joint angular characteristics on the accuracy of industrial robots. Industrial Robot: An International Journal 38, 406–418 (2011)
17. Erkaya, S.: Investigation of joint clearance effects on welding robot manipulators. Robotics and Computer-Integrated Manufacturing 28, 449–457 (2012)

18. Gong, C., Yuan, J., Ni, J.: Nongeometric error identification and compensation for robotic system by inverse calibration. International Journal of Machine Tools and Manufacture 40, 2119–2137 (2000)
19. Ruderman, M., Hoffmann, F., Bertram, T.: Modeling and Identification of Elastic Robot Joints With Hysteresis and Backlash. IEEE Transactions on Industrial Electronics 56, 3840–3847 (2009)
20. Kumagai, S., Ohishi, K., Miyazaki, T.: High Performance Robot Motion Control Based on Zero Phase Error Notch Filter and D-PD Control. In: IEEE International Conference on Mechatronics, Malaga, Spain, pp. 1–6 (2009)
21. Marton, L., Lantos, B.: Friction and backlash measurement and identification method for robotic arms. In: IEEE International Conference on Advanced Robotics, pp. 1–6 (2009)
22. Thomsen, S., Fuchs, F.W.: Speed Control of Torsional Drive Systems with Backlash. In: Proceedings of 13th European Conference on Power Electronics and Applications, pp. 1–10 (2009)
23. Carvalho Bittencourt, A., Wernholt, E., Sander-Tavallaey, S., Brogardh, T.: An Extended Friction Model to capture Load and Temperature effects in Robot Joints. In: IEEE International Conference on Intelligent Robots and Systems, Taipei, pp. 6161–6167 (2010)
24. Jin, M., Jin, Y., Chang, P.H., Choi, C.: High-Accuracy Trajectory Tracking of Industrial Robot Manipulators Using Time Delay Estimation and Terminal Sliding Mode. In: Proceedings of 35th Annual Conference of IEEE Industrial Electronics, pp. 3095–3099 (2009)
25. Merlet, J.P.: Interval analysis for certified numerical solution of problems in robotics. International Journal of Applied Mathematics and Computer Science 19, 399–412 (2009)
26. Zhang, H., Wang, J., Zhang, G., Gan, Z., Pan, Z., Cui, H., Zhu, Z.: Machining with Flexible Manipulator: Toward Improving Robotic Machining Performance. In: Proceedings of IEEE International Conference on Advanced Intelligent Mechatronics, Monterey, California, USA, pp. 1127–1132 (2005)
27. Zhang, H., Pan, Z.: Robotic Machining: Material Removal Rate Control with a Flexible Manipulator. In: Proceedings of IEEE Conference on Robotics, Automation and Mechatronics, pp. 30–35 (2008)
28. Pa, Z., Zengxi, P., Hui, Z., Zhub, Z., Wanga, J.: Chatter analysis of robotic machining process. Journal of materials processing technology 173, 301–309 (2006)
29. Quintana, G., Ciurana, J.: Chatter in machining processes: A review. International Journal of Machine Tools and Manufacture 51, 363–376 (2011)
30. Liu, X.-W., Cheng, K., Webb, D., Longstaf, A.P., Widiyarto, M.H.: Improved dynamic cutting force model in peripheral milling. Part II: experimental verification and prediction. International Journal of Advanced Manufacturing Technology 24, 794–805 (2004)

Machining with Industrial Robots:
The COMET Project Approach

Christian Lehmann[1], Marcello Pellicciari[2], Manuel Drust[3],
and Jan Willem Gunnink[4]

[1] Chair of Automation Technology, Brandenburg University of Technology,
Siemens-Halske-Ring 14, 03046 Cottbus, Germany
[2] DIEF Engineering Department Enzo Ferrari, University of Modena and Reggio Emilia,
Via Vignolese 905/B, 41125 Modena, Italy
[3] Fraunhofer Institute for Manufacturing Engineering and Automation IPA,
Nobelstrasse 12, 70569 Stuttgart, Germany
[4] Delcam PLC Small Heath Business Park, Birmingham, B10 0HJ. UK
lehmann.christian@tu-cottbus.de,
pellicciari.marcello@unimore.it,
manuel.drust@ipa.fraunhofer.de, jwg@delcam.com

Abstract. Machining using industrial robots is currently limited to applications with low geometrical accuracies and soft materials due to weaknesses of the robot structure, insufficient controller performance and the lack of suitable software tools. This paper proposes a modular approach to overcome these obstacles, applied both during program generation (offline) and execution (online). Offline predictive machining errors compensation is achieved by means of an innovative programming system, based on kinematic and dynamic robot models. Realtime adaptive machining error compensation is also provided by sensing the real robot positions with an innovative tracking system and corrective feedback to both the robot and an additional high dynamic compensation mechanism on piezo-actuator basis. Due to the modularity of the approach, an individual setup can be compiled for each actual use-case. Final experimental validation of the components is currently ongoing in multiple robot cells, covering several application areas as aerospace, automotive or mould construction.

Keywords: robotic machining, engineering methods, 3D-piezo-actuator compensation mechanism.

1 Introduction

The wide and extensive use of industrial robotics is finally leveraging smart manufacturing. Industrial robots intrinsic re-configurability and adaptiveness is in fact crucial to cope with the latest requirements of extreme responsiveness and flexibility in production [1] [2] and enabled many successful new applications. Nowadays, industrial robots are effectively used to perform complete industrial processes integrating intelligent manipulation, manufacturing and quality control.

P. Neto and A.P. Moreira (Eds.): WRSM 2013, CCIS 371, pp. 27–36, 2013.

However, despite the huge potential applications, machining with industrial robots is still an unsolved problem and the only real applications are de facto limited to less demanding tasks with (very) low accuracy and low material removal rate, mainly for deburring, deflashing and finishing purposes [3].

The COMET project [4] addressed the robot machining challenge and developed a modular and configurable platform able to enhance the machining accuracy of standard industrial robots enabling cost-effective, first time right, robot machining. In this paper it will be presented the COMET project approach to the problem, the solutions developed and a brief outlook on the results obtained.

Standard industrial robots have strong performance limitations that have precluded their use for machining tasks. In the COMET project such limitations were initially carefully investigated and used as foundation to formulate the design requirements.

The main challenges arise from the machining process itself: in fact machining is a process intrinsically dynamic and time varying where high motion accuracy must be assured in presence of elevated, and continuously changing, forces and disturbances. On the other hand, industrial robots are extremely rationalized (i.e.: cost-effective), well engineered products, designed for high repeatability and not accuracy: as a matter of fact it is overall known that their intrinsic compliant structure under working loads leads to important positional errors, usually compensated by teaching points coordinates different from the nominal ones. The main limits to overcome for machining with industrial robots are then related to their weak motion accuracy in presence of high and continuously variable process loads. The main sources of loss of motion accuracy can be briefly summarized as:

- **Robot overall mechanical stiffness** (including joint compliance): on average industrial robots have a stiffness less than 1 N/μm, much lower respect to the one of standard CNC machine tool centers, often greater than 50 N/μm [5]. Moreover, it is important to remember that the robot overall stiffness is strictly dependent from its configuration [6].
- **Robot joints accuracy**: the adoption of high ratio gear reducers leads to high, non linear and load dependent, frictions losses and, most important, the inevitable presence of backlash (on catalogue data [7], the reducer by itself claims around 1', that cannot be neglected within the wide robot working envelopes), which however is rarely addressed, when optimizing robot accuracy for machining applications.
- **Robot real geometry**: well evaluated by kinematic calibration.
- **Robot control system**: robot controllers must face important limits compared to the corresponding CNC machine tools ones. In fact, the robot controller must deal with a flexible manipulator whose base natural frequency, usually from around 20Hz up to 10Hz with large payloads on the TCP, is much lower than the one of a CNC machine tool, which are at least several hundred Hz [5]. Furthermore, state of the art robot controllers memory is not able to manage large amount of data and the paths taught are usually much less accurate than the ones programmed in CNC machine tools.
- **Process forces**: as already written, machining process forces are not negligible due to the limited robot stiffness while their continuous unpredictable variations lead to chatter and overall vibrations.

In order to assess the influence of these motion accuracy errors sources on the final machining quality, Design of Experiments driven machining tests were executed on several robot cells. The results clearly showed the joints backlash as the major source of machining accuracy errors while robot controllers showed their limits, mainly in terms of absolute accuracy and integration with external realtime feedbacks loops. The joint backlash by itself was evaluated to lead machining errors up to 0.5mm, with a primary influence from axis 1 together with 2 & 3.

Since the main limits and sources of errors in robot machining were identified, the research focus was then oriented to solve the most important ones.

2   The COMET Project Approach to Robot Machining

An effective industry oriented robot machining requires proper error compensation solutions able to overcome the intrinsic performance limitations of standard industrial robots respect to machine tools. The COMET project approach is focused on a novel modular, and configurable, machining error compensation platform that can be customized for specific application fields with different accuracy and performance.

2.1   Basic Concept

In the COMET project two different adaptive error compensation approaches were developed: offline compensation, based on the predictive calculation of the robot motion accuracy errors and their consequent correction, and online compensation, based on the realtime measurement of the real robot TCP position for an active compensation action. Such two different approaches are based on four main modules, which address the issues described in Section 1:

1) A unique Kinematic and Dynamic representation of each Industrial Robot entity (KDMIR), including a methodology to determine the respective required model parameters. The respective modeling and parameter identification procedures are separately described in [8] and [9].

2) An integrated Programming and Simulation environment for adaptive generation of the machining path for Industrial Robots (PSIR), which builds upon the unique robot models. The implemented mechanisms for inclusion of different (robot) models into the CAM environment are discussed in [10].

3) An Adaptive Tracking system for Industrial Robots (ATIR), which detects deviations from the desired robot path and initiates realtime corrective actions towards the robot controller.

4) A High Dynamic Compensation Mechanism (HDCM) which can perform additional positional corrections that exceed the robots mechanical bandwidth or its positional accuracy. The mechanism follows the idea of a 3D-piezo compensation mechanism previously presented in [11].

By combining these modules (summarized in Figure 1), different configurations for the setup of the industrial robot machining cell are possible. The first important distinction has to be made between predictive error compensation applied offline during programming (KDMIR and PSIR) and the realtime compensation applied online during machining (ATIR and HDCM). Again for each subdivisions can be made, depending if certain sub-modules are integrated or not (e.g. online compensation only with the robot itself or with robot and HDCM together). These will be explained further in the Sections 2.2 and 2.3. With such modular platform it is possible to configure the robot cell optimizing the performance for a specific application. The cell layout can be designed with configurations where the robot moves the milling spindle or the workpiece. Furthermore, the overall approach by principle is of general use and robot vendor independent.

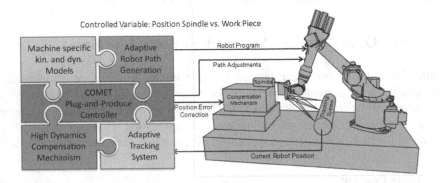

Fig. 1. Schematic summary of the COMET modules and overall approach

2.2   Offline Compensation

The COMET project Programming and Simulation environment for Industrial Robots (PSIR) aims at realizing a complete, first time right, robot machining path program, avoiding the need of long and complex commissioning on the real robot cell.

The outcome of the developed Programming and Simulation environment for Industrial Robots (PSIR) should be a complete and correct robot path, which does not require changes to be applied within the robot cell. This is an important requirement due to the usage for machining applications, where only an initial tool and work piece calibration is possible (as also on regular CNC machines). Manual corrections of further points are not possible due to the huge number of tool path points. Work piece based learning is often not acceptable due the long machining times and the high costs per work piece. Therefore the software needs to consider possible issues beforehand and either correct them directly or display them to the user for manual correction. In order to improve the achievable accuracy, additionally the kinematic and dynamic deviations of the robot structure have to be considered. The approach described in the following therefore aims at modeling and compensating the machining error sources during robot program generation. The robot path is adapted according to the foreseen deviations. So that the robot is not commanded to the desired pose, but to a pose that will end near the desired one, after all errors affected the robot arm. To foresee the positional errors of the robot, the robot is modeled with components reflecting the mechanical issues as described above, namely the optimized kinematic description and a coupled model of the robot joints, including backlash, friction and torsional stiffness for each joint. In order to utilize this robot model for compensation of the machining path, some additional information from the process itself is required to determine the forces active on the tool during machining. Therefore additional to the updated kinematic description and the joint-based robot model, a model to estimate the process forces is required, which again needs detailed info from the CAM system about the material and tool as well as the engagement situation of the tool in order to give a valid output. Starting point of the offline compensation is a tool path defined within the CAM system. In contradiction to conventional machine tools this tool path – besides information about tool position and orientation – due to the additional degree-of-freedom, also includes information about the respective robot poses. The subsequent chain of applied calculations after such a tool path has been generated is the following:

- Within the CAM system an engagement angle calculation is executed in order to predict the engagement situation of the tool within the material for each point of the tool path.
- Based on the predicted engagement situation of the tool, a 3D process force vector is calculated, predicting the magnitude and direction of the force working on the tool tip (for more information on the force calculation based on KIENZLE [12] the reader is referred to [8]). The force calculation considers both the machined material as well as the tool geometry. The calculated force is the force affecting the robot, either directly (if the robot is moving the spindle) or indirectly as reaction force (if the robot is moving the work piece).
- With the combined information on how the robot should move according to the CAM and which forces affect the tool (and thus the robot) an external simulation using the robot model on joint basis (as described in [4]) can be run, first determining how the robot would actually move due to the joint based effects (like backlash and friction in the gears or compliance of the joints) and consecutively generate altered program points to compensate for these effects.
- In a last step a kinematic calibration is applied, which again alters the points of the tool path, using both the nominal kinematic values (which also the robot controller uses internally) and optimized parameters based on measurements (which better reflect the actual kinematic structure).

At the end a regular robot program is generated, using a post-processor for the respectively used robot brand. As all compensations are done by adapting the cartesian points in the robot program and no additional information has to be transferred towards the robot, application of this approach is independent of the robot brand. Disadvantage of the compensation per program point is the hereby limited resolution (see [10]). Resolution enhancement is only possible up to certain limits, determined e.g. by controller memory or cycle time. Although internally the separate simulations can be run with higher resolutions, but for the outputted program these limits persist. The described process chain cannot only be used to generate compensated robot paths, but alternatively can also be used to simulate the behaviour of the robot *without compensation*. In combination with a material removal simulation the machining outcome when using the uncompensated robot can be visualised in order to determine potential geometrical errors and the overall achievable machining accuracy.

In order to apply the offline compensation on a certain robot, different model parameters have to be determined first, which are then stored in a so-called *Robot Signature file* (this file is created for each unique robot and can be accessed by the CAM system to load the respective model parameters). Different measurement and parameter identification methods have been developed or utilized. The optimized kinematic parameters are identified using an optical tracking system and measurements of the end effector in free space movements. For the determination of the joint based parameters an identification method for kinematic parameters [13] (based on the idea of generating a closed kinematic chain by rigidly clamping the robot to the ground) was applied to the identification of joint properties [14], [9]. For the identification of the material and tool dependent parameters a method which processes force data captured during machining of a test work piece was developed [8].

2.3   Online Compensation

This part proposes an approach for the online error compensation in the range of micrometers during machining tasks of industrial robots. The concept takes into account data acquisition, sophisticated data fusion and external compensation using a parallel 3D-piezo-actuator compensation mechanism (HDCM). In this case the robot positions the workpiece relative to the tool. The tool is mounted on the HDCM which allows the adjustment of the tool in the working range of the HDCM.

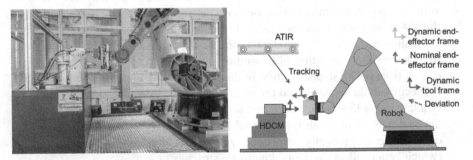

Fig. 2. (a) Experimental set-up for online compensation at Fraunhofer IPA, (b) Measured frames of the set-up.

Considering the compensation idea there are two mechanical systems for compensation. Firstly, the robot which is comparably slow, but has a large workspace. Secondly, the HDCM which demands conversely to the robot fast movements in a limited geometric working range. As a result the deviation between nominal end-effector frame and dynamic end-effector frame relative to the dynamic tool frame needs to be adjusted. This determines the control-error. To fully understand the set-up the deviation is depicted in Figure 2. The online measurements are obtained by using a metrological tracking system. Therefore path deflections of the robot e.g. generated by the backlash or compliance can be measured. For instance the Nikon Metrology (COMET project partner) K600 system allows dynamic tracking of two frames at the same time. Beside dynamic measurements static measurements are initially necessary to match the set-up with the actual set-up. One could obtain values of static frames from CAD. But in practice, despite accurate construction static errors in the set-up are expected. This is resolved by doing a cell calibration based on the usage of metrological tracking system. Taking into account that the HDCM is designed for fast but small compensations, the saturation of each axis of the HDCM has to be avoided. Therefore the determined error is partitioned between robot and HDCM. As robot and HDCM behaves differently smart splitting between both systems is introduced. This approach realizes a frequency-partition of the control-error in low-frequency errors to be compensated by the robot itself and the high-frequency errors to be compensated by the compensation mechanism HDCM.

In order to fulfill the criteria of high dynamic compensation a progressive design is implemented based on the experiences described in [15]. As shown in [11] piezo-actuators combined with ESSJs-lever-mechanisms are appropriate for smooth movements. Opposed to conventional bearings friction, play and backlash are significantly reduced. The chosen approach uses instead of a serial mechanism a parallel actuation to improve the dynamical behavior. The reduction of the moved mass allows improving the dynamics. Furthermore, the real achievable working range depends strongly on the stiffness of the transmission system. In particular the piezo actuator only allows actuating up to a certain force, because of its bounded stiffness. The additional load influences therefore the working range. Piezo-actuators are equipped with strain gauge sensors. Additional capacitive sensors are placed underneath the movable plate. Those capacitive sensors are aligned with the axes directions. Thus, each capacitive sensor allows the tracking of the movable plate in one compensation direction. A feedback control approach realizes the automatized positioning of each axis, respectively. Input voltage is set deforming the piezo-actuator. The proposed control scheme for each compensation axis takes into account:

- Inner PID controller for feedback from strain gauge sensors in piezo-actuators for handling parameter uncertainties and disturbances
- Outer model based controller for position control of the end-effector plate where the machining spindle is attached. The control of a piezo-actuated high-dynamic compensation mechanism is presented in [16] and [17]. The control variable is measured by the capacitive sensors.

3   Cell Setups and Demo Strategy

Experimental validation was made in a total of eight demonstration setups, covering different basic setups (spindle on robot or on fixture), different robot brands (ABB, KUKA, YASKAWA Motoman) and different application cases. Depending on the requirements of the respective demo application, a specific set of the modules as described in Section 2.1 was applied, with the PSIR as basic component for all cells. Applications range from aero and automotive components (complete machining and finishing processes for aluminium and inconel parts) to mould making (complete machining in hardened steel, requiring high accuracies). First tests were made with simplified geometries on test work pieces (see Fig. 3a) to show the general feasibility to machine the requested materials and to validate the developed compensation modules. Depending on the complexity of the compensation, this validation was carried out in several subsequent steps, f.i. for the offline compensation the different calculation steps (as described in Sec. 2.2) were first tested on their own before validating the complete compensation chain. Secondly the developed methods and components were used to machine industrial parts from the various industrial sectors, highlighting the combination possibilities of the modular approach. In Fig. 3b such a demonstration part (with rough machining on the sides and semi-finishing in the middle) made from hardened steel (X37CrMoV5-1) is shown. For the shown demo part (depending on applied compensation) a geometrical accuracy of 0.4 mm or below can be achieved.

Although the COMET approach is designed robot brand independent, certain restrictions for some robot cells limited the applicability of the developed compensation methods, either due to missing access to (controller) values required during the identification of the model parameters or as consequence of missing possibilities to feed back correctional values for the online compensation in a sufficient quality.

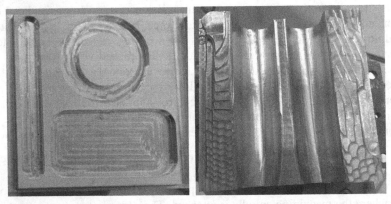

Fig. 3. (a) Test features used for validation of the different compensations, (b) Machining of industrial demonstration parts (mould-and-die)

The machining of the industrial parts showed that machining with industrial robots can be an alternative to the use of dedicated machine tools, although the actual benefits are clearly depending on the specific use-case and material. For soft materials machining results comparable to machine tools are possible, but also for more challenging materials like hardened steel, robots can be a viable alternative. The applied compensations allow manufacturing within tolerances which are sufficient for roughing and semi-finishing for hard materials, so that capacities can be taken off the (costly) machine tools for these steps where their high (final) accuracy is not necessarily required. The biggest remaining geometrical deviations occur where process conditions change rapidly (e.g. for material entry or exit). For less demanding materials complete machining of the industrial parts is possible.

Another conclusion that can be drawn is that – besides the improvements possible with compensations – a fair amount of quality can be gained already by selection of appropriate machining strategies. Not only that a proper strategy can enhance the work piece quality already on its own, the resulting – more stable and predictable – cutting process offers a much more reliable basis for application of the compensations.

4   General Observations

From the machining experiments so far it can be concluded that, besides the improvements that can be achieved using the different proposed compensation methods, also the general cell setup and the selected machining strategy have an important influence both on the achievable geometrical accuracy and the resulting surface quality. Ensuring stable cutting conditions is the foundation for reliable application of the compensation approach. While the general proposed approach is robot brand independent, the implementation at the demo cells showed differences in applicability depending on the respective brand but also differences between different types of the same brand. The heterogenic situation on the robot market therefore still is an obstacle for each solution aiming at improving robot machining accuracy.

Further work is required on the combination of the compensations applied offline and online. Up to this point only one group of compensations can be used at once. Obstacles here are the need to transfer both the compensated and the nominal path throughout the whole process chain, as well as the synchronization between the different representations of the machining process in general (ideal and actual movements and forces) and the tool path in particular (point-based in the robot program, but required time-based for the online compensation).

Finally, future works may address the deeper integration with robot vendor's controllers for real-time feedback loops through external sensors, actually felt as the main performance limit. Furthermore, the analysis of the ongoing tests final results will stimulate future development guidelines on the COMET approach basis.

Acknowledgments. The research work reported here was supported by the European Commission under the Seventh Framework Programme (FP7/2007-2013) within the project COMET under grant agreement #258769.

References

1. Tolio, T., Ceglarek, D., ElMaraghy, H.A., Fischer, A., Hu, S.J., Laperrière, L., Newman, S.T., Váncza, J.: SPECIES—Co-evolution of products, processes and production systems. CIRP Annals - Manufacturing Technology 59(2), 672–693 (2010) ISSN 0007-8506, 10.1016/j.cirp.2010.05.008
2. Pellicciari, M., Leali, F., Andrisano, A.O., Pini, F.: Enhancing Changeability of Automotive Hybrid Reconfigurable Systems in Digital Environments. International Journal on Interactive Design and Manufacturing 6, 251–263 (2012)
3. Surdilovic, C., Dragoljub, Zhao, H., Schreck, G., Krueger, J.: Advanced methods for small batch robotic machining of hard materials. In: Proc. of ROBOTIK 2012: 7th German Conference on Robotics, Munich, Germany, May 21-22, pp. 1–6 (2012)
4. COMET project, http://www.comet-project.eu/ (accessed December 18, 2012)

5. Pan, Z., Zhang, H., Zhu, Z., Wang, J.: Chatter analysis of robotic machining process. Journal of Materials Processing Technology 173(3), 301–309 (2006)

6. Pan, Z., Zhang, H.: Robotic machining from programming to process control: a complete solution by force control. Industrial Robot: An International Journal 35(5), 400–409 (2008)

7. Seiki, T.: http://www.tkkcorporation.com/nabtesco/nabtesco.htm (accessed January 17, 2013)

8. Lehmann, C., Halbauer, M., Euhus, D., Overbeck, D.: Milling with industrial robots: Strategies to reduce and compensate process force induced accuracy influences. In: Proc. of 17th IEEE International Conference on Emerging Technologies & Factory Automation, ETFA 2012, Kraków, Poland (2012)

9. Lehmann, C., Olofsson, B., Nilsson, K., Halbauer, M., Haage, M., Robertsson, A., Sörnmo, O., Berger, U.: Robot Joint Modeling and Parameter Identification Using the Clamping Method. In: Proc. of IFAC Conference on Manufacturing Modelling, Management and Control, MIM 2013, Saint Petersburg, Russia (2013)

10. Lehmann, C., Halbauer, M., van der Zwaag, J., Schneider, U.: Offline Path Compensation to Improve Accuracy of Industrial Robots for Machining Applications. In: Proc. of 14th Automation Congress, Baden-Baden, Germany (2013)

11. Puzik, A.: Genauigkeitssteigerung bei der spanenden Bearbeitung mit Industrierobotern durch Fehlerkompensation mit 3D Ausgleichsaktorik, Dissertation, University of Stuttgart, Fraunhofer IPA (2011)

12. Kienzle, O.: Bestimmung von Kräften an Werkzeugmaschinen. VDI-Z 94, 299–305 (1952)

13. Bennett, D., Hollerbach, J., Henri, P.: Kinematic calibration by direct estimation of the Jacobian matrix. In: Proc. of IEEE Int. Conf. on Robotics and Automation, ICRA, Nice, France, pp. 351–357 (1992)

14. Nilsson, K.: Patent Application SE-1251196-0: Method and System for Determination of at Least One Property of a Manipulator (2012)

15. Puzik, A., Meyer, C., Verl, A.: Industrial Robots for Machining Processes in Combination with an 3D-Piezo-Compensation-Mechanism. In: 7th CIRP International Conference on Intelligent Computation in Manufacturing Engineering, CIRP ICME 2010: Innovative and Cognitive Production Technology and Systems, Capri, Italy, June 23-25 (2010)

16. Olofsson, B., Sornmo, O., Schneider, U., Robertsson, A., Puzik, A., Johansson, R.: Modeling and control of a piezo-actuated high-dynamic compensation mechanism for industrial robots. In: Proc. of 2011 IEEE/RSJ International Conference on Intelligent Robots and Systems, IROS, San Francisco, USA, September 25-30, pp. 4704–4709 (2011)

17. Sornmo, O., Olofsson, B., Schneider, U., Robertsson, A., Johansson, R.: Increasing the milling accuracy for industrial robots using a piezo-actuated high-dynamic micro manipulator. In: Proc. of 2012 IEEE/ASME International Conference on Advanced Intelligent Mechatronics, AIM, Kaohsiung, Taiwan, July 11-14, pp. 104–110 (2012)

A Calibration Method for the Integrated Design of Finishing Robotic Workcells in the Aerospace Industry

Francesco Leali, Marcello Pellicciari, Fabio Pini,
Alberto Vergnano, and Giovanni Berselli

"Enzo Ferrari" Engineering Department, University of Modena and Reggio Emilia,
via Vignolese 905/B, 41125 Modena, Italy
{francesco.leali,marcello.pellicciari,
fabio.pini,alberto.vergnano,giovanni.berselli}@unimore.it

Abstract. Industrial robotics provides high flexibility and reconfigurability, cost effectiveness and user friendly programming for many applications but still lacks in accuracy. An effective workcell calibration reduces the errors in robotic manufacturing and contributes to extend the use of industrial robots to perform high quality finishing of complex parts in the aerospace industry. A novel workcell calibration method is embedded in an integrated design framework for an in-depth exploitation of CAD-based simulation and offline programming. The method is composed of two steps: a first offline calibration of the workpiece-independent elements in the workcell layout and a final automated online calibration of workpiece-dependent elements. The method is finally applied to a robotic workcell for finishing aluminum housings of aerospace gear transmissions, characterized by complex and non-repetitive shapes, and by severe dimensional and geometrical accuracy demands. Experimental results demonstrate enhanced performances of the robotic workcell and improved final quality of the housings.

Keywords: Workcell Calibration, Industrial Robotics, Integrated design, Aerospace industry.

1 Introduction

Most mechanical components in aerospace industry are characterized by complex shapes and narrow tolerance ranges, to accomplish light weight design requirements and to comply with standards and safety regulations.

The manufacturing process generally consists of two machining steps. Starting from a cast part, the first step involves the CNC machining of the functional features. In the second step the part is manually finished by removing the burrs remaining from previous machining or in zones of difficult reachability. Manual finishing requires highly skilled operators but does not assure a constant dimensional and geometric quality and it is extremely time-consuming [1].

Manufacturing robotic workcells generally provide a performing and economically sustainable alternative to manual finishing. Robotic changeable automation, in particular, has been demonstrating to be an effective solution due to its

P. Neto and A.P. Moreira (Eds.): WRSM 2013, CCIS 371, pp. 37–48, 2013.

(re)configurability, (re)programmability and relative low cost. Its changeability, defined as the ability to cope with change or uncertainty, represents a key factor for small lot production of complex shaped parts [2].

In a changeable robotic workcell, the system is capable of heavy changes in its layout, in the settings of the mechatronic devices and in the robot's program blocks, depending on the product characteristics or on the technological process performed. The set-up time at changeover can be reduced through a modular architecture, developed following tailored design methods [3].

The authors already developed and proposed a method for designing reconfigurable robotic workcells, based on a PPR approach (Product analysis, Process identification and Resource selection). The integrated use of CAD-based environments and offline programming (OLP) tools is fundamental to satisfy the requirements on system changeability and to easily reduce the time lost for robot's reprogramming at changeover [4], [5].

Extending the application field of industrial robots from simple deburring to complex finishing and machining, the main operative limitation is given by their low precision in posing, defined in terms of resolution, repeatability and accuracy [6], [7]. The most important values used to represent precision performances of manipulators, as specified in the international standard ISO 9283, which sets the performance criteria of industrial manipulators, are pose repeatability and pose accuracy [8].

Kinematic and dynamic performances of robotic arms, in particular, depend both on mechanical and control factors and represent the major contributions to the final pose repeatability and accuracy [9].

Manufacturing and assembly tolerances on robot links introduce variations in their dimensions while the robot controller, set with nominal values, does not consider the own parameter variability of single manufactured robots.

Other typical mechanical errors, affecting the robot kinematic and dynamic behavior, are backlashes on gear and belt transmissions, friction on harmonic drives and bearings, and the intrinsic low stiffness of the robotic mechanical chain, around 1N/mm, with respect to conventional CNC tool machines, with stiffness greater than 50N/mm [6].

Besides, the difference between the actual and the physical joint zero configuration set in the robot controller represents another importance source of uncertainties for an accurate robot pose definition.

Dynamic errors mainly depend on servo system accuracy, encoder resolution, system inertia and friction, so the robot controller is finally responsible for the trajectory deviation from its nominal definition, also due to physical loads acting on the robots (e.g. payload, gravity).

During finishing and machining, contact forces between tools and workpieces, for both part-in-hand and tool-in-hand robot configurations, influence the part quality in terms of dimensions and geometries. The tool dynamic behavior has to be investigated and machining parameters have to be carefully chosen to minimize robot's chattering and structure deformations and unsteadiness [10].

The design of auxiliary equipment is another important factor for achieving high quality in finishing and machining. Devices for tool wear control and part manipulation, for instance, need to be manufactured to assure exact reference workframes. Finally, the control of environmental factors as temperature and working conditions (swarf and dust collection, etc.) is very important to minimize the machining errors [11]. Fig.1 left side summarizes the main error sources affecting the robotic machining process.

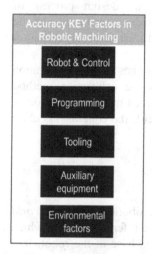 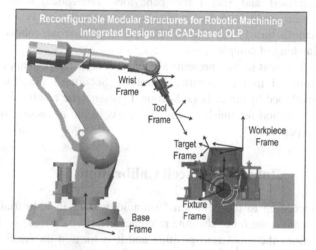

Fig. 1. Accuracy factors in robotic machining (left side) and main reference frames in a reconfigurable workcell – case with spindle on robot (right side)

In robotic finishing and machining, more than in other operations, programming and code generation represent demanding tasks. In fact the robot has to run tailored approaches with hundreds of target points toward parts subject to geometric and dimensional variability and generally inaccurately posed.

In reconfigurable workcells, moreover, the robot code has to be designed to be modular and easy to be (re)used. The robot code subprograms, workpiece dependent, have to be collected and adapted at every cycle, while non-dependent instructions have to be carefully parameterized.

Manual programming is widely diffused in robotic finishing and machining to comply with many difficulties, so that a lot of time is wasted at changeover, leading to a notable productivity loss.

OLP software, integrated by specific machining functions or dedicated packages, currently offered by several robot manufacturers and software companies for simulation and code generation, have demonstrated the capability to reduce the productivity loss at changeover [12].

The full correspondence between the real and the virtual controller implemented in the software guarantees an exact analysis of the robot's behavior and suggests a possible partial correction of the most important kinematic and dynamic errors [13].

On the other hand, the OLP approach, when applied to the design of reconfigurable robotic workcells, introduces a group of reference frames which must be used by skilled programmers, expert both in manufacturing and robot programming (for each robot brand) to modularize the robot program architecture and also define the nominal position of each element which takes part to the robotic process. Modules of robot programs and nominal position of element commonly correspond to specific reference frames, as shown in Fig.1. Anyway, misalignments between nominal and real positions of such elements could frustrate the advantages given by OLP in enhancing the robot's accuracy.

The present paper proposes a calibration method to bridge the gap between simulated and real robot behaviors, embedded within the design process of reconfigurable robotic workcells. The final goal is to reduce errors in robotic manufacturing and extend the use of industrial robots to perform high quality finishing of complex parts.

The next section presents a brief review of the main approaches in robot calibration proposed in the scientific literature. Section 3 describes the calibration method developed by the authors. Section 4 presents the results obtained by the application of the method for finishing complex parts in the aerospace field; the conclusions close the paper.

2   Robotic Workcell Calibration

According to the common terminology used in industrial robotics [14], the *World Frame* is the robot absolute reference. The *Base Frame* is the reference system which defines the robot zero position and it is located on the fixed base of the robot. In a workcell with one robot only, the *Base Frame* matches with the *World Frame*. The *Wrist Frame* is located on the robot wrist flange and defines the robot kinematic chain. Objects attached to the robot flange are referenced with specific frames; in case of spindle moved by the robot a *Tool Frame* defines the position of every tool tip. The *Fixture Frame* is used to identify and locate stationary or movable fixtures with respect to the *Base Frame*. The *Workpiece Frame* defines the relation between the workpiece zero point and its relative fixture, while a *Target Frame* is used to describe the robot's configuration at every point of a work path. The robot movements are defined with a sorted sequence of matches between the *Tool Frames* and the *Target Frames*.

As already introduced, an inaccurate definition of the reference frames has a direct impact on the final accuracy. The robot's kinematic and dynamic errors strictly depend on the correspondence between the nominal and the real dimensions of every robot link, also referred as *absolute positioning inaccuracy*. The misalignment from nominal to real poses of each element which takes part to the manufacturing action in the workcell layout contributes to the *relative positioning error* [14], [15].

The workcell calibration is the action of matching the nominal and the real pose of every reference frame which enters into the definition of the whole robotic manufacturing cycle. So a full workcell calibration considers both the singular parts of the robot (absolute calibration) and the poses of the workpieces, tools and mechatronic devices involved in finishing and machining (relative calibration) [14], [15].

Many research efforts have been spent in the past to define absolute calibration methods and instruments. The most common absolute methods are the so called model-based parametric and non-parametric calibrations [16].

Model-based calibration improves the robot accuracy through a parametric identification of the main physical error sources. Model-based calibration is defined by [17] and [18] along three main levels: *Joint, Kinematic* and *Nonkinematic*. The Joint Level calibration corrects the relationship between the signal produced by the transducer at every joint and its actual displacement, involving drives and joints' sensors. The Kinematic level calibration acts on the entire robot kinematic model, aiming at determining the basic kinematic geometry of the robot as well as the correct angles of every joint. The last level covers errors in robot positioning due to the dynamic effects, such as joint compliance, friction and clearance, and link compliance.

Non-parametric calibration estimates the robot's positioning errors through analytical interpolation methods which start from the measurement of the mechanical properties of the robot in predefined configurations. Absolute instruments as laser trackers or 3D cameras can be effectively employed for robot absolute calibration. Some examples can be found respectively in [13], [16], [19].

The relative positioning errors are mainly caused by dynamic changes and drift of the mechanical and/or electronic devices during the robotic operations and are difficult to correct. Other error sources are related to dimensional and geometric variations of the elements, misalignments due to the workpiece feeding and pose, tool wearing and installation of devices [15].

A further classification in relative calibration is introduced by [20], where the authors propose the distinction between calibrations which adopt external instruments and calibrations based on the use of the robot itself. Examples of external devices are coordinate measuring systems, which can be used to measure the position of the devices within the robotic workcell. This approach is particularly time consuming and has to be applied every time the workcell configuration changes, very frequent in case of small production batches. On the contrary the robot itself can be used as a carrier for accurate measuring sensors. State of the art instruments and procedures are described in [21], [22], [23].

Concluding, relative calibration highly depends on the specific robotic workcell configuration and manufacturing application so few methods are proposed as general solution for the issue. In [14] another method is presented, but it has still to be fitted into an integrated design loop involving CAD-based simulation and OLP tools.

3   Calibration Method

In order to deliver high quality robotic machining, an original method is presented to fully integrate the calibration approach into the design loop of reconfigurable workcells.

As previously described, calibration compares the real poses of the equipment in the robot workcell against the reference frames defined in the OLP environment. The workcell equipment can be classified in two main categories: workpiece-independent and workpiece-dependent elements. Workpiece-independent elements must be selected in function of the robotic process operations, and define the basic structure of the reconfigurable workcells. On the other hand, the workpiece-dependent elements are specific for the singular workpiece or workpiece family. Positioners, robots and tool racks are examples of workpiece-independent elements while workpiece fixtures and machining tools represent workpiece-dependent elements [24].

According to these two categories, the workcell calibration includes two different steps. First a layout workcell calibration is performed, defining the relative position between the robot and the workpiece-independent elements. This approach is well known in industry and allows to correct the errors share due to the mechanical inaccuracy. It is repeated just one time after the workcell assembly. Manufacturing performances obtained following this first calibration step are however not good enough to satisfy the demanding requirements of high quality finishing and machining.

Then an online calibration is realized to calibrate the reference frames defined for tools, workpiece-dependent elements and for the workpieces themselves. Such step is repeated every time a new workpiece starts its finishing or machining cycle.

The proposed calibration method uses the robot itself as a measuring machine, since the robot *Wrist Frame* defines the spatial position of the tool flange. To effectively exploit a robot for measuring, two extra sensors are adopted. A stationary sensor calibrates the reference frame of the robot end-effectors while a robot on-board sensor calibrates the reference frames of the elements not handled by the robot. The calibration process follows four main steps, as outlined in Fig.2.

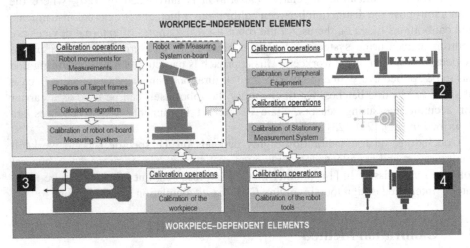

Fig. 2. Equipment and operations involved in the calibration method

The step 1 carries out the *Calibration of the robot on-board measuring system*. At this stage the robot is employed to define the reference system of the robot on-board sensor. The robot is moved along predefined directions with respect to a fixed accurate calibration element located in the workcell. Whenever the sensor recognizes the reference element, a target frame is self-learned by the robot. A calibration algorithm processes the recorded positions and calculates the accurate robot position relative to the on-board sensor reference frame.

The step 2 involves the robot and on-board sensor to realize the *Calibration of the peripheral equipment* and the *Calibration of the stationary measuring system*, following the same calibration process previously described. Steps 1 and 2 accomplish the layout workcell calibration one time only, after the workcell installation.

At step 3 of the proposed method, the procedure *Calibration of the workpiece* defines the workpiece frame, exploiting the robot and its on-board sensor. The sensor locates the workpiece with respect to given references features, like workpiece holes, pins or planes. The *Calibration of the robot tool*, step 4, defines the tool frames location, and is performed using the stationary sensor located within the work volume of the robot. The last two steps conclude the workcell calibration, and are performed for each single processed workpiece.

Following the approach already proposed by [5], the method is implemented through the definition of parametric modules which simultaneously embed geometric characteristics, control logics and robot code for a quick simulation in a CAD-based offline programming environment, optimization and automation of the calibration process itself.

4   Robotic Workcell Calibration for Improving Finishing Quality in Aerospace Industry

The proposed method has been employed for the integrated design of a robotic workcell for accurate finishing of aerospace components. Workpieces are part of a family of aluminum housings for gear transmissions with an envelope from 4x10-3m3 to 8x10-2m3. Fig.3 shows one housing and a detail before (right side, on the top) and after the robotic finishing (right side, on the bottom).

Fig. 3. Workpiece and a detail before (right top) and after finishing (right bottom)

Fig.4 shows the workcell layout. The element #1 represents the workcell iron floor, designed for a quick transportation of the robotic workcell and to univocally fix the workpiece-independent elements. The industrial robot selected for the process, identified by #2, is an ABB IRB 2400/16 with 16 kg payload and enhanced stiffness thanks to the parallelogram linkage on the third link. The workpiece feeding system at

#3 is an indexed positioner with two controlled tilting frames which are used to orient the workpiece during the finishing cycle. Each workpiece is locked on a mechanical interface for its quick positioning on the fixtures (element #4). The workpiece posing and blocking are realized through a customized set of reference pins, as shown at #5. The robot mounts a change system for a quick replacements of the end-effectors (e.g. the compliant pneumatic spindle #6), stored in the rack #7. The stationary sensor #8 is a fork light barrier, located near the tool rack in order to reduce the time needed for the online calibration of the robot tools. A Renishaw RP1 inspection probe equipped with a ruby ball stylus is the on-board sensor used for both the initial calibration of the workpiece-independent elements and for the online measuring of the workpiece itself.

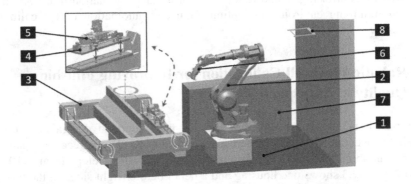

Fig. 4. Reconfigurable robotic workcell layout in ABB RobotStudio environment

ABB RobotStudio 5.14.03 is the CAD-based OLP tool adopted to implement the calibration method within the integrated design loop.

Advanced digital models of the measuring sensors have been developed through the smart functions provided by the OLP environment, for simulating, optimizing and commissioning the calibration procedure defined by the method.

A changeable and open architecture is then developed to replicate the logic behaviour of the sensors, linking independent parametric functional modules to a common skeleton, as proposed in [4].

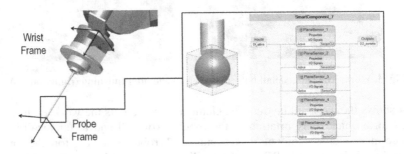

Fig. 5. The advanced model developed to replicate the logic behavior of the touch probe

Focusing on the touch probe shown in Fig.5, five planar collision detection features replicate the probe behavior during the touching action. I/Os gates are used to communicate with the robot controller and stop the robot motors in event of collision (probe deflection). The model contains also the subprogram for robot programming which can be easily recalled within the digital environment and reused for further simulations.

The workpiece frame calibration procedure calculates the·frame pose relying on measurements of a set of *Target Frame* at a planar surface and along two holes on the workpiece through a list of commands previously set and simulated by the user in the OLP environment. Thus, the virtual controller runs the complete calibration process, computing and updates the robot programming data type which defines the frame location, called *wobjdata* in the ABB RAPID native language.

Fig.6 shows the offline simulation of the calibration process (on the left), the online calibration of the workpiece reference frame by the on-board sensor (at the center) and the online calibration of a finishing disc though the aforementioned stationary sensor (on the right), particularly important to correct the errors due to the tool wear.

Fig. 6. Offline programming (left), online workpiece (center) and tool calibration (right)

In Fig.7 two measuring results on the workpiece borders machined with the robotic approach confirm that the border sections satisfy the quality specification. Such limit corresponds, for the application considered, to the minimum value among 0.726mm and 0.3 times the length of the shortest nearest edge.

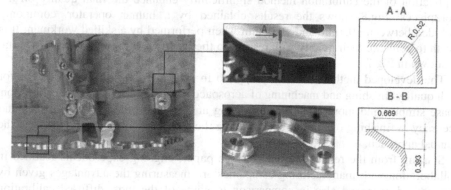

Fig. 7. Finishing of the gear transmission housing

5   Conclusions and Future Work

The aerospace industry fulfills small lots production of parts that require finishing processes with high accuracy. Reconfigurable robotic systems could satisfy these requirements but need integrated design methods to enhance their modularity, both for the mechatronic devices and robot programs. Nevertheless, industrial robots deliver insufficient kinematics and dynamics accuracies for finishing operations, so it is essential to correct the different error sources.

The proposed approach leads to the development of a virtual prototype where the main features are referred to reference frames linked to every modular workcell element. The reference adjustment to match real and nominal positions can be realized with respect to the robot (absolute calibration) or to the relative positions between the workcell elements (relative calibration).

The article proposes a new calibration method consisting of two steps. First the robot handles a measuring touch probe and records every workcell element positions with respect to the spatial *World Frame*. This approach is already usually executed in industrial workcells just after their installation.

In the second step the robot, equipped by an on-board sensor, online measures the workpiece reference frame and the workpiece-dependent elements while a stationary sensor is used for tool reference frame online calibration.

The major novelty is that the method fully integrates the calibration phase within the engineering design cycle, i.e. finishing strategy validation, process simulation, offline programming and robot path generation, robot code commissioning. This is possible since the virtual prototype integrates the dimensional and geometric information of the used devices, the control logic necessary for the interaction with the mechatronic systems and the robot code.

The offline programming of the finishing paths and the following online calibration provide the accuracy required for the workpiece finishing, i.e. the breakage of machined to machined intersections. The final machining accuracy is a sum of various contributions; pose repeatability and accuracy, for instance, for an ABB IRB 2400/16 is ±0.03mm [25].

Calibration does not impact on all the values of the error chain. Nevertheless the application of the calibration method significantly enhance the final quality on the workpieces. Fig.8 shows the results obtained by a human operator, commonly included between ±0.1mm and ±0.05mm when performed by a skilled workman. It is worth to compare their quality with respect to the quality achieved by the robotic arm, as shown in Fig.7.

The developed method is actually applied in programming three robotic cells for high quality finishing and machining of aerospace parts, since it makes the calibration phase simpler and more effective, improving at the same time the final quality and accuracy in finishing operations and reducing the loss productivity with respect to the manual approach.

Starting from the results presented in the paper, ongoing experiments realized in collaboration with manufacturing companies are measuring the advantages given by the method proposed also in comparison to some of the most diffused calibration methods adopted in Industry.

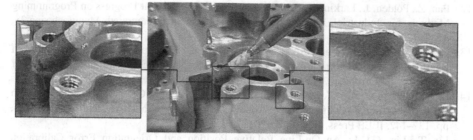

Fig. 8. Manual finishing of the housing (center) and details of the machined edges (left and right)

Acknowledgments. The authors want to express their gratitude to Luciano Passoni, Davide Passoni and Lino Ferrari from SIR S.p.A. (Modena, Italy), for their technical and managerial contribution to the project, and AVIO S.p.A. (Torino, Italy) for supporting the experimental tests.

References

1. Jayaweera, N., Webb, P.: Robotic Edge Profiling of Complex Components. Industrial Robot: An International Journal 38, 38–47 (2010)
2. ElMaraghy, H.A.: Flexible and Reconfigurable Manufacturing Systems Paradigms. International Journal of Flexible Manufacturing Systems 17, 261–276 (2005)
3. Bi, Z.M., Lang, S.Y.T.: General Modular Robot Architecture and Configuration Design. In: 2005 IEEE International Conference on Mechatronics and Automation, pp. 268–273. IEEE Press, New York (2001)
4. Pellicciari, M., Leali, F., Andrisano, A.O., Pini, F.: Enhancing Changeability of Automotive Hybrid Reconfigurable Systems in Digital Environments. International Journal on Interactive Design and Manufacturing 6, 251–263 (2012)
5. Andrisano, A.O., Leali, F., Pellicciari, M., Pini, F., Vergnano, A.: Hybrid Reconfigurable System Design and Optimization Through Virtual Prototyping and Digital Manufacturing Tools. International Journal on Interactive Design and Manufacturing 6, 17–27 (2012)
6. Pan, Z., Zhang, H.: Robotic machining from a programming to a process control: a complete solution by force control. Industrial Robot: An International Journal 35, 400–409 (2008)
7. DePree, J., Gesswein, C.: Robotics Machining White Paper Project. Halcyon Development - Robotic Industries Association (RIA)
8. Abderrahim, M., Khamis, A., Garrido, S., Moreno, L.: Accuracy and Calibration Issues of Industrial Manipulators. In: Huat, L.K. (ed.) Industrial Robotics: Programming, Simulation and Application, pp. 131–146. Pro Literatur Verlag, Germany (2006)
9. Shiakolas, P.S., Conrad, K.L., Yih, T.C.: On the Accuracy, Repeatability, and Degree of Influence of Kinematics Parameters for Industrial Robots. International Journal of Modelling and Simulation 22, 1–10 (2002)
10. Pan, Z., Zhang, H., Zhu, Z., Wang, J.: Chatter Analysis of Robotic Machining Process. Journal of Material Processing Technology 173, 301–309 (2006)
11. Cherif, M., Knevez, J.Y., Ballu, A.: Thermal Aspects on Robot Machining Accuracy. In: Proceedings of IDMME – Virtual Concep. (2010)

12. Pan, Z., Polden, J., Larkin, N., Van Duin, S., Norrish, J.: Recent Progress on Programming Methods for Industrial Robots. Robotics and Computer-Integrated Manufacturing 28, 87–94 (2012)
13. COMET Project – Plug-and-Produce COmponents and METhods for Adaptive Control of Industrial Robots Enabling Cost Effective, High Precision Manufacturing in Factories of the Future. In: European 7th Framework Programme, reference number 258769
14. Li, X., Zhang, B.: Toward General Industrial Robot Cell Calibration. In: 5th IEEE International Conference on Robotics, Automation and Mechatronics (RAM), pp. 137–142. IEEE Press, New York (2011)
15. Lu, T., Lin, G.C.I.: An On-Line Relative Position and Orientation Error Calibration Methodology for Workcell Robot Operations. Robotics and Computer-Integrated Manufacturing 13, 89–99 (1997)
16. Lim, H.K., Kim, D.H., Kim, S.R., Kang, H.J.: A Practical Approach to Enhance Positioning Accuracy for Industrial Robots. In: ICCAS-SICE 2009 ICROS-SICE International Joint Conference 2009, pp. 2268–2273. IEEE Press, New York (2009)
17. Mooring, B.W., Roth, Z.S., Driels, M.R.: Fundamentals of Manipulator Calibration. John Wiley & Sons, Inc., New York (1991)
18. Elatta, A.Y., Li, P.G., Fan, L.Z., Yu, D.Y.: An Overview of Robot Calibration. Information Technology Journal 3, 74–78 (2004)
19. Mustafa, S.K., Pey, Y.T., Guilin, Y., I-Ming, C.: A geometrical Approach for Online Error Compensation of Industrial Manipulators. In: IEEE/ASME 2010 International Conference on Advance Intelligent Mechatronics, pp. 738–743. IEEE Press, New York (2010)
20. Gan, Z., Sun, Y., Tang, Q.: In-Process Relative Robot Workcell Calibration. US Patent 6,812,665 (November 2, 2004)
21. Cheng, F.S.: The Method of Recovering Robot TCP Positions in Industrial Robot Application Programs. In: 2007 IEEE International Conference on Mechatronics and Automation (ICMA 2007), pp. 805–810. IEEE Press, New York (2007)
22. Bley, H., Bernerdi, M., Franke, C., Seel, U.: Process-based assembly planning using a simulation system with cell calibration. In: 2001 IEEE International Symposium on Assembly and Task Planning, pp. 116–121. IEEE Press, New York (2001)
23. Ribeiro, F., McMaster, R.: A Low Cost Cell Calibration Technique and its PC based Control Software. In: 1997 IEEE International Symposium on Industrial Electronics (ISIE 1997), vol. 3, pp. 840–845. IEEE Press, New York (1997)
24. Andrisano, A.O., Leali, F., Pellicciari, M., Vergnano, A.: Engineering Method for Adaptive Manufacturing Systems Design. International Journal on Interactive Design and Manufacturing 3, 81–91 (2009)
25. ABB Product Specification Articulated Robot IRB 2400 – 3HAC9112-1, Rev. T

Pneumatic Driven Device for Integration into Robotic Finishing Applications

Paulo Abreu and Manuel Rodrigues Quintas

IDMEC – Pólo Feup, Faculdade de Engenharia,
Universidade do Porto, 4200-465 Porto, Portugal
{pabreu,mrq}@fe.up.pt

Abstract. The use of robots in industrial applications has been widespread from handling tasks to processes. The finishing processes include operations such as deburring, grinding and polishing. Within these processes, there is a need to control the contact force between the workpiece and the robot. The force control can be implemented in a passive or active form, either in the robot arm or with an external device. This paper presents the development of a device to be used in robotic finishing applications. It provides a semi-active system that limits and maintains the contact force and is used in conjunction with the robot controller. The device uses a pneumatic driven linear axis fitted with a position sensor. The architecture of the developed system and experimental results regarding the performance of the built prototype are presented. The implementation of different control strategies to adjust the robot path velocity are proposed and discussed.

Keywords: Robotic grinding, compliant tool holder.

1   Introduction

The use of robots in industrial applications has been widespread from handling tasks to processes such as spot welding, mig/mag welding, painting, sealing and finishing operations. The finishing processes typically carried out by industrial robots include operations such as deburring, deflashing, grinding, polishing and machining [1], [2]. Within these processes, the problems resulting from the contact forces between the workpiece and the robot are of major importance not only to the mechanical integrity of the robot, but also for the success of the task. As such, the most used robotic finishing operations are grinding and polishing that do not require very high contact forces. It is therefore important to provide force control [3]. The force control can be implemented in a passive or active form, either in the robot arm or through the use of an external device [4]. As industrial robots are typically position/velocity controlled, the use of external devices is widely employed to control/limit the contact forces [5], [6]. There are several compliant tools to be used by robots or CNC machines that can provide passive or active compliance, making use of pneumatic actuators [7], [8], [9]. These tools normally are able to provide linear compliance, radial compliance or a combination of both. When used in conjunction with a robot, normally the robot is

P. Neto and A.P. Moreira (Eds.): WRSM 2013, CCIS 371, pp. 49–56, 2013.

programmed to place the workpiece along a predefined path at a given velocity and the tool is allowed to deflect along its compliance direction ensuring a constant contact force. Linear compliance tools offer high stiffness in the perpendicular direction of the control force (path direction), while radial compliance tools have lower stiffness in the path direction [4].

When using active force control with robots, it is used a force sensor to measure the contact force so that the robot programmed path trajectory or velocity can be adjusted in real-time [10]. Examples of such robotic systems are the ones supplied by different robot manufactures as reported in [11]. Although these systems offer higher accuracy and repeatability than passive devices, they are very expensive and the industrial world has been slow to use them. ABB is one of the robots manufactures that offers active force control [12]. The robot is equipped with a force sensor from ATI and the controller (IRC5), with the software RobotWare Machining FC, provides two working modes, FC Pressure and FC Speed Change. With FC Pressure, the programmer specifies the direction and value of the desired applied force and the robot will adjust the programmed path in order to keep the reference force. This working mode is best suited for polishing/grinding applications and can accommodate slightly variation of parts size. With FC Speed Change, the programmer specifies the paths, velocities and the contact force that, when reached, will slow down the execution of the path. If the force decreases, the robot will return to the previous path velocity. This working mode is appropriated for material removal/deburring tasks, enabling to achieve a lower cycle time to process a given workpiece. Based on this working mode, it was envisaged the development of a low cost compliance device for integration with a robotic cell to be used in deburring tasks [13]. The device holds the deburring tool providing an adjustable compliance through the use of a pneumatic actuator, while the robot, holding the workpiece, is under position/velocity control. Furthermore, the device is able to communicate with the robot controller, sending the position error of the tool; the robot is then programed to change the path velocity based on the toll position error.

2   Development of the Device

In a robotic grinding application, it is normally required that the robot executes a given path while maintaining the contact between the tool and the workpiece, because it is necessary to achieve a given geometry for the workpiece. As the robot is controlled under position/velocity control, any deviations from the programmed path result in an increase of the contact force or even in the absence of contact. This contact force is required to be kept within a given value to assure that the tool can accomplish the material removal. Normally, if the workpiece is light, the robot holds the workpiece and the tool is in a fixed position. In this type of application the contact area between the tool and the workpiece is normally small. In many cases it is possible to consider that the contact occurs in a point or line. In such cases, to assure the perpendicularity between the workpiece surface and the tool, the robot assures the correct position of the workpiece. This allows the tool to be mounted on a single axis of movement, simplifying the construction of such a tool holder so that springs or pneumatic actuators can be used to control/limit the contact force.

The developed system is based in an architecture where the robot moves the workpiece while the tool is fixed to the compliance device, as shown in Fig. 1. The compliance of the device is assured by a linear pneumatic actuator with both chambers under adjusted pressure. The device has one degree of freedom provided by a linear table and its relative position can be measured. An adjustable mechanical stop is provided to define the working position of the tool. This feature is of particular importance, since the moving table is kept in contact with this mechanical stop. The displacement from this position, that occurs only when the robot reaches the defined contact force, is sent to the robot controller and, in the present case, used to change the robot path velocity. Since the robot has the flexibility to locate the workpiece in any given position and orientation, it is considered that the compliance device can only be provided with one degree of freedom. It is also considered that this solution is sufficient to cover the requirements of grinding processes, thus facilitating the mechanical design and reducing production costs.

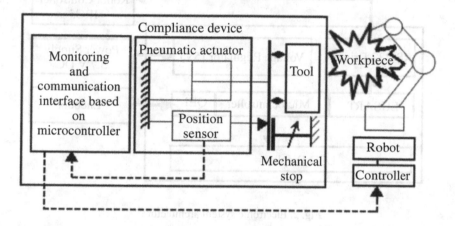

Fig. 1. Architecture of the integrated system

The compliance device is built around a linear table from Rexroth Star (model SGO 12 – 85) that assures the linear movement required. This linear axis has a travel length of 165 mm, low friction and a maximum load capacity of 1500 N and maximum load torque of 52 Nm and 57 Nm in motion direction and in orthogonal motion direction, respectively. The table holds the tool. To power the movement of the table it is used a pneumatic double acting cylinder. It is used a low friction cylinder from SMC (MQMLB10-60D) that has a total travel length of 60 mm. With this cylinder it is possible to apply a theoretical force within the range of 0.05 N – 55 N, depending on the supply pressure (0-0.7 MPa) and on the precision of the pressure regulation. The pressure regulator valves are from SMC (model SMC IR1020-01). These are precision valves, with 0.2% sensitivity and a repeatability of +/- 0.5% in relation to the applied pressure (Maximum pressure of 1.0 MPa). The digital pressure indicators are from SMC (model ISEA30A-01).   The position of the table is measured with an incremental encoder. The encoder has a tape with a resolution of

150 LPI (lines per inch). As the signal is read in quadrature, it is possible to achieve a linear resolution of 0.042 mm. To measure the displacement of the tool, monitor the position and communicate the measured position to the robot controller, it was developed an electronic system based on a microcontroller from Microchip running at 8 MHz (model PIC18F2431). The architecture of the electronic system is presented in Fig. 2. This microcontroller has an interface for serial communications EUSART (Enhanced Universal Synchronous Asynchronous Receiver Transmitter), which is used to communicate with the robot controller through the RS232 port, with a baud rate of 19200 bps. It has also an interface to connect to an incremental encoder (QEI - Quadrature Encoder Interface) with the capability to measure position and velocity. To program the microcontroller it was used a development kit - EasyPIC 4, from Mikroelektronika.

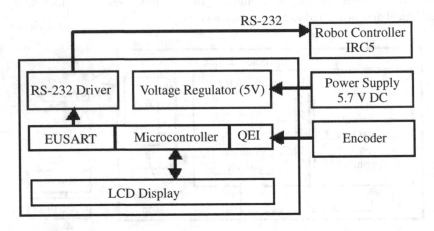

Fig. 2. Electronic system architecture

To setup the RS232 communication, that uses a voltage signal within the range [-10 V, +10 V], and due to the fact that the microcontroller is powered with a 5 voltage supply, it was necessary to use an integrated circuit (MAX232 from Maxim) to accommodate this difference. The position of the carriage can be expressed in 14 bits. This data is codified in two bytes using binary operations at the level of the microprocessor programming. The high byte has its most significant bit set to one and the low byte has its most significant bit set to zero. This process enables to transfer the position information to the robot controller in an efficient way, saving memory space and assuring that the controller is able to correctly identify the two bytes. The electronic system is provided with a reset button that is used to reset the counter used to read the pulses from the encoder. To monitor the position of the encoder, in real-time, the electronic system has a LCD display. A picture of the developed device with the electronic system is presented in Fig. 3.

Fig. 3. Pneumatic device

3 Experimental Results

To verify the device is able to impose a constant force when the tool is moved from its working position, the robot was programmed to execute a linear movement, at a controlled velocity against the tool. The force sensor the robot has mounted on its end effector was used to measure the real contact force. The programmed trajectory of the robot is presented in Fig. 4 and comprises four linear movements. These movements were performed at two distinct velocities - 10 mm/s and 50 mm/s. The pressure regulators were adjusted to provide two nominal contact forces, 9 N and 25 N. The values of the contact force were recorded every 0.1s. The results are presented in Fig. 4 for the test performed with the robot moving at 10 mm/s and with a nominal contact force set to 25 N. The device was able to keep the contact force within the programmed reference force, with an error of 2-3 N. This error is due to the friction forces in the guiding system, in the cylinder and in the pneumatic circuit. For the test performed at the same velocity but with a lower contact force (9 N), the behaviour was found to be similar.

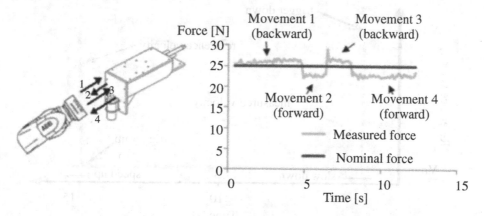

Fig. 4. Force response with robot velocity of 10mm/s and nominal force of 25 N

To implement the adjustment of the robot path velocity based on the displacement of the tool from its working position, two different control strategies were used. These control strategies are implemented within the program that runs on the robot controller. The first one uses a continuous linear reference velocity while the second one uses a discrete two steps reference velocity with hysteresis (Fig. 5).

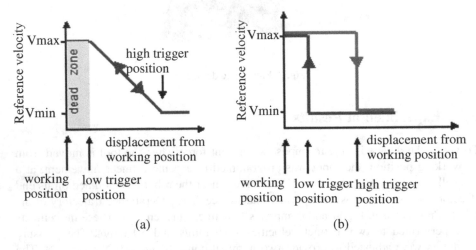

(a)   (b)

Fig. 5. Control strategies

For the first control strategy, it is defined a dead zone corresponding to an initial displacement of the tool, where the robot controller receives a constant reference velocity. When the tool moves outside this position (the low trigger position is reached), the reference velocity starts to decrease linearly (Fig. 5 a). To adjust the

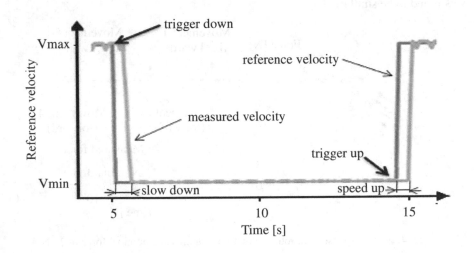

Fig. 6. Experimental result in finishing operation

robot path velocity, based on the measured position of the tool, the robot is programmed using the instruction SpeedRefresh from the robot programming language (RAPID). It was verified that the robot presented a response time of 0.4s to update its velocity. This delay, coupled with the continuous generation of reference velocities by the controller, reduces the dynamic response of the system in updating the robot path velocity. This lead to adopt the second control strategy that uses only two reference velocities with hysteresis (Fig. 5 b).

To test this strategy it was conducted an experiment in finishing a linear wood ruler with a pneumatic grinding tool. The robot was programmed with two reference velocities (100 mm/s and 10mm/s). The hysteresis width was 0,25mm and the low trigger position was set to 1.25mm. The reference contact force was set to 9 N. The data was registered at a rate of 0.1s. The test results regarding the control parameters are presented in Fig. 6.

It can be seen the robot slows down when the tool reaches the programmed trigger position (high) that occurred due to the presence of excessive "burr". When the tool returns to the programmed trigger position (low), the robot will accelerate to the programmed velocity. This control strategy presented a better global dynamic behaviour, although the delay in the update of the robot velocity was kept within the 0.4 s. This fact is due to the use of the instruction SpeedRefresh provided by the robot programming language to implement the adjustment of the robot velocity, since the data communication between the microcontroller and the robot controller do not introduce any significant delay.

Based on these findings it is envisaged to extend the use of the developed device to adjust the speed of the tool to be implemented with the microcontroller. The position of the tool will then be used to adjust not only the robot path velocity but also the speed of the tool. With this new control strategy, it will be possible to adjust the material removal rate, based on the relative velocity between the tool and the workpiece, defined by the robot path velocity and the tool velocity.

4   Conclusions

This paper presents the development of a device to be used in robotic finishing applications. This device provides a semi-active system that limits and maintains the contact force and is used in conjunction with the robot controller. The device uses a pneumatic driven linear axis fitted with a position sensor. Experimental results regarding the performance of the built prototype and its integration with the robot were presented. The device was able to keep the contact force within the programmed reference force, with an error of 2-3 N. Two different control strategies that use the position of the tool to adjust the robot path velocity were proposed. The strategy that uses a commutation between two reference velocity levels with hysteresis presented better performance. The need to implement this control strategy within the robot controller and with the available programming functions is responsible for a delay of 0.4s in the velocity update. The developed system, with this control strategy, is able to implement finishing tasks where the material removal rate can be controlled based on the adjustment of the robot path velocity, while maintaining a constant contact force.

References

1. Graf, T.: Deburring, Finishing, and Grinding Using Robots and Fixed Automation: Methods and Applications. In: International Robots & Vision Automation Conference, pp. 20.1–20.22 (1993)
2. Bogue, R.: Finishing robots: a review of technologies and applications. Industrial Robot: An International Journal 36(1), 6–12 (2009)
3. Erlbacher, E.A.: Force control basics. Industrial Robot: An International Journal 27(1), 20–29 (2000)
4. Odham, A.: Active and Passive Force Control Robotic Deburring. In: RIA Grinding, Deburring, and Finishing Workshop, St. Paul, Minnesota (2007)
5. Liao, L., Xi, F., Liu, K.: Modeling and control of automated polishing/deburring process using a dual-purpose compliant toolhead. International Journal of Machine Tools and Manufacture 48(12-13), 1454–1463 (2008)
6. Kim, C., Chung, J.H., Hong, D.: Coordination control of an active pneumatic deburring tool. Robotics and Computer-Integrated Manufacturing 24(3), 462–471 (2008)
7. ATI Industrial Automation, http://www.ati-ia.com
8. PushCorp, Inc., http://www.pushcorp.com
9. RAD, The Robotic Accessories Leader, http://www.rad-ra.com/Deburring-Tools.htm
10. Song, H.C., Song, J.B.: Precision Robotic Deburring Based on Force Control for Arbitrarily Shaped Workpiece Using CAD Model Matching. International Journal of Precision Engineering and Manufacturing 14(1), 85–91 (2013)
11. Marvel, J., Falco, J.: Best Practices and Performance Metrics Using Force Control for Robotic Assembly. NIST publications (2012), http://dx.doi.org/10.6028/NIST.IR.7901
12. Fixell, P., et al.: A touching movement. ABB Review 4, 22–25 (2007)
13. Viana, D.G.: Desenvolvimento de uma solução robótica para operações de acabamento de solas de sapatos. MSc Thesis, Faculty of Engineering, University of Porto, Portugal (2010)

An Agile Manufacturing System
for Large Workspace Applications

Hai Yang[1], Cédric Baradat[1], Sébastien Krut[2], and François Pierrot[2]

[1] Industry and Transport Division, Tecnalia France
672, rue du Mas de Verchant, Montpellier 34000, France
{hai.yang,cedric.baradat}@tecnalia.com
[2] Robotics Department, LIRMM CNRS
161, rue Ada, Montpellier 34000, France
{sebastien.krut,francois.pierrot}@lirmm.fr

Abstract. REMORA aims at offering an agile robotic solution for manufacturing tasks done on very large parts (e.g.: very long and slender parts found in aeronautic industries). For such tasks, classical machine-tools are designed at several tens of meters. Both their construction and operation require huge infrastructure supports. REMORA is a novel lightweight concept and flexible robotic solution that combines the ability of walking and manufacturing. The robot is a mobile manufacturing system which can effectuate operations with good payload capacity and good precisions for large workspace applications. This new concept combines parallel kinematics to ensure high stiffness but low inertia, and mobile robotics to operate in very large workspaces. This results in a machining center of new generation: 1. Agile manufacturing system for large workspace applications; 2. Heavy load and good precisions; 3. 5-axis machining and 5-axis locomotion/clamping; 4. Self-reconfigurable for specific tasks (workspace and force); 5. Flexible and multifunctional (machining, fixtures…).

Keywords: Agile manufacturing, Mobile manufacturing, Parallel Kinematics, Redundancy.

1 Introduction

Modern industries and services request larger and larger pieces of equipment: aircraft spars, pipelines sections, wind turbine wings and hubs, etc. The increasing size of such equipment requires the appropriate means of manufacture, which sets great challenges for the machine-tool industry. For these tasks, existing solutions are at their limits [1]. Stationary arms suffer from their limited workspaces. Manipulators mounted on vehicles are often not accurate and rigid enough. Classical machine-tools must be designed at mega-scale (several tens of meters). In order to introduce mobile robotic solutions to certain large workpiece manufacturing processes, one has to consider problems linked to: locomotion, stiffness and localization in large workspaces.

P. Neto and A.P. Moreira (Eds.): WRSM 2013, CCIS 371, pp. 57–70, 2013.
© Springer-Verlag Berlin Heidelberg 2013

Parallel kinematic mechanisms (PKMs) which have a great potential to provide high rigidity and motion dynamics suffer from an inherent limitation in their operational workspace. Legged robots have attracted attention because of their relatively good terrain crossover capacity. Although, numerous works on legged robots have led to remarkable improvements on various aspects such as mobility [2], dynamics equilibrium [3], payload capacity [4] etc., most of these robots are designed for exploration or military purposes. Few of them have been used to solve industrial problems due to control complexity and lack of reliability [5].

The development reported in this paper is related to an innovative robot which combines the ability to walk on the workpiece (or on the tooling that supports the workpiece) together with manufacturing ability. The mechanical concept is based on a redundancy parallel mechanism: eight motors for six degrees-of-freedoms (DoFs). Combining motors, brakes, clamping devices and numerous position sensors, the robot can clamp itself on a tooling for the manufacturing tasks, and then it can change its configuration to become a walking robot able to reach the next working area. The overall system and the function modes of the robot are presented firstly, and then the issues of redundancy are briefly addressed. Based on the built prototype, a typical step of the robot is illustrated. In the end, the advantages and potential applications as well as future works are discussed.

2   System Description

REMORA is an industry-oriented legged/parallel robot which has several specific features such as lockers on certain joints, clamping device on the extremity of each limb. Combining with proper control strategies, the robot switches between different working modes for achieving manufacturing tasks in large workspaces.

2.1   Overall Description of Robot

The system is composed of the robot itself and the clamping pins fixed on the supporting media which can be the floor, the fixture of workpieces or other structures in the workshop (Fig. 1). The CAD Model of the robot is shown in Fig. 2. Clutches

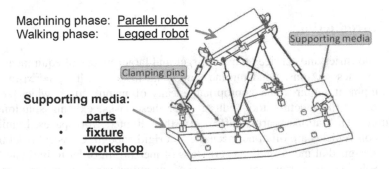

Fig. 1. General description of the system

are fitted on several passive joints of the robot. Furthermore, there is a clamping device on the extremity of each limb, which provides climbing capacity to the robot.

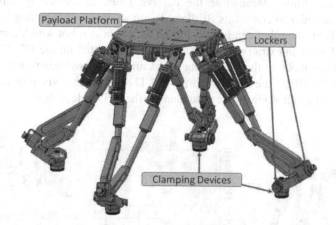

Fig. 2. CAD of the robot

2.2 Working Mode

The robot works in two kinds of working modes:

Payload Platform (PP) mode: As illustrated in Fig. 3, every limb of the robot is attached to the supporting media. Each limb possesses six DoFs; two of them are actuated. The PP of the robot, which has six DoFs, is actuated by eight linear jacks located in the four limbs connected to the base. Therefore, in this mode the robot can be considered as a six DoFs PKM with actuation redundancy.

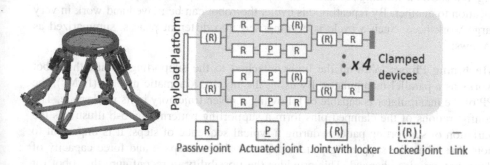

Fig. 3. PP Working Mode - Parallel robot (6-dof platform; 8 actuators)

Branch Extremity (BE) mode: One limb of the robot is detached from the base in order to reach another clamping pin. The other three limbs remain attached to the base (Fig. 4). The PP is still fully controllable in 6 DoFs using the actuators located in the three attached limbs. Meanwhile the passive DoFs in the swing limb should be eliminated in order to control its BE link. To do so, the corresponding lockers will be activated. Then the robot can be considered as a hybrid robot which consists of a six DoFs PKM with an extra two DoFs end-effector mounted on the PP. There are eight actuators, both those in the supporting limbs and in the swinging limb, contribute to positioning the BE of the swinging limb in 3-D space with a given orientation. Thus, the robot can be called kinematic redundant in the BE mode.

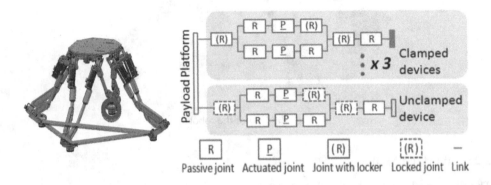

Fig. 4. BE Working Mode - Legged robot (6-dof platform, 6 actuators + 2-dof arm, 2 actuators = 8-dof walking, 8 actuators)

2.3   Working Scenario

Fig. 5-a shows a working scenario which presents one operation cycle from one work location to another. By repeating this cycle, the robot can be moved and work in very large workspace. Such a cycle is decomposed into different phases, summarized as follows:

Machining Phase: With all the limbs attached to the supporting points, the robot works as a parallel manipulator. By using the inverse kinematic models (IKMs), the PP of the manipulator is capable of following a given trajectory in its workspace. The configurations of the clamped pins form a supporting pattern. Fig. 5-b illustrates the variation of supporting patterns during a typical sequence of steps. It is important to note that when the supporting pattern changes, the workspace and force capacity of the robot are also changed. This provides the possibility to reconfigure the robot for various tasks.

Reconfiguration for Limb Swinging Phase: During this phase, all the limbs of the robot are still attached to the supporting points. Before locking the corresponding joints, the PP will move to a specific position with a given pose in order to have all the required lockable joints in the desired positions for locking. Properly choosing the locking poses is one of the key issues for ensuring the robot to reach the desired next clamping pin.

Limb Swinging Phase: With the corresponding lockers activated, the extremity of the swinging limb can follow a given 6-axis trajectory to the next supporting point. Such swinging phase is divided into three sub-phases: Extraction of the limbs, Free swinging and Insertion of the limbs.

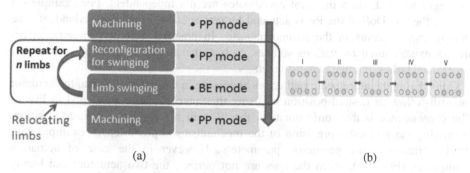

(a)   (b)

Fig. 5. (a) Working scenario; (b) Typical sequence of steps

3   Redundant Issues

As it has been mentioned in the previous section, there are two kinds of redundancies appearing in the structure during the different working modes: the actuation redundancy for the PP mode and the kinematic redundancy for the BE mode, respectively. Appropriate management of the redundancies is one of the key issues for ensuring a good realization of tasks.

3.1   Actuation Redundancy

Actuation redundancy is an interesting feature for improving the workspace properties of parallel robots: (1) it helps to build PKMs with larger singularity-free workspace; (2) it also helps to homogenize the force performances over the whole workspace [6].

However the control of PKMs with actuation redundancy can be a source of problems. The PID control loops shown in Fig. 6 is the most implemented controller

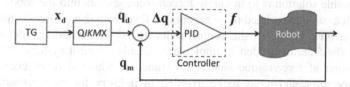

Fig. 6. Conventional control scheme

in most industrial applications. T.G. stands for Trajectory generator, QIKMX is the inverse kinematics model of PKMs with the pose variable $\mathbf{x}$ as the input and the actuator variable $\mathbf{q}$ as output, PID is the proportional, integral and derivative control block, $\mathbf{x_d}$ is the desired pose variable of the PP, $\mathbf{q_d}$ is the desired actuator positions, $\mathbf{q_m}$ is the measured actuator positions, Δ_q is the difference between the desired actuator positions and the measured actuator positions, the actuators forces can be considered as proportional to $\mathbf{f}$, the input of actuator amplifiers are set in torque mode.

When there is actuation redundancy, such a classical PID joint space controller is no longer adapted, since the joint coordinates are not independent. For example (in Fig. 7), the one DoF of the PP is actuated by one actuator in the non-redundant case and by two actuators in the redundant case. In practice, certain geometric errors always exist, caused by various sources such as machining inaccuracies, assembly errors, backlashes, thermal expansion, etc.

Considering the aforementioned errors appearing in the bar in Fig. 7-a, the actuator can still follow its desired position by using the control scheme presented in Fig. 6. The consequence is that, unfortunately, the PP will not reach the desired position. Improving the geometric precision of the mechanisms is possible, for example with the calibration of the geometric parameters. However in the case of actuation redundancy (Fig. 7-b), when the bars are not perfect, the two actuators can barely reach the desired positions at the same time. By applying the classical control

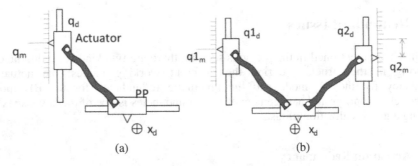

(a)   (b)

Fig. 7. (a) Conventional bar; (b) Redundancy bars

scheme in Fig. 6, the two actuators will continually 'fight against' each other, which will generate undesired internal forces. Such forces cause unnecessary energy consumption, and with the effect of integral terms in the PID controller, it may even be destructive for the mechanism.

One possible solution is to integrate force/torque sensors into the robot, so that the internal force can be regulated to an acceptable level with some force/torque feedback control strategies. However, the control units of industrial robots sometimes do not present all the features needed to implement complex control systems. Furthermore, the integration of force/torque sensing in standard industrial robot control units is often cumbersome and tends to be avoided in industry for many reasons such as reliability, cost, etc.

In order to treat properly the actuation redundancy, we apply an innovative joint space PID controller which is illustrated in Fig. 8.

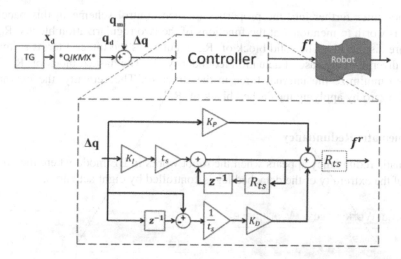

Fig. 8. Proposed control scheme for mechanisms with actuation redundancy

Based on the standard PID controller, two regularization matrix blocks are introduced to the control scheme. One possible way to construct the regularization matrix is presented as follows.

Regularization is carried out after the PID block as illustrated in Fig. 8 where f is regularized in the force space according to the following steps:

1. The wrench in the Cartesian space is calculated according to the forces in the actuation space thanks to the force mapping relationship.
2. Such a wrench in the Cartesian space is then used for calculating the new regularized actuation forces by the inverse force mapping relationship.

The two steps can be written in one equation:

$$f^r = \left(J_m^T\right)^+ J_m^T f \tag{1}$$

Where f represents the actuation forces directly calculated by the PID block according to Δ_q (unregularized differences of the joint positions), f^r represents the regularized actuation forces which are not supposed to generate any internal forces in the mechanism. J_m is the inverse kinematic Jacobean Matrix. As the mechanism has actuation redundancy, the J_m should not be considered as the inverse matrix of kinematic Jacobian Matrix which is not invertible in this case. It is obtained directly from the static kinetostatic relationship.

Accordingly, the regularization matrix R_{ts} can be defined as:

$$R_{ts} = \left(J_m^T \right)^+ J_m^T \tag{2}$$

We will not enter further into the properties of such control scheme in this paper. Instead, it is worth to mention that the functions of the two regularization blocks 'R_{ts}' in Fig. 8 are distinct. 1.) The solid block of 'R_{ts}' regularizes the integral terms so that it will not diverge in the case of actuation redundancy. 2.) The dashed block of 'R_{ts}' is supposed to minimize the internal force in the structure. Theoretically, the internal force can be zero by applying the dashed block of 'R_{ts}'.

3.2   Kinematic Redundancy

The kinematic redundancy appears when the robot works in BE mode where the pose (6-Dofs) of the extremity of the detached leg is controlled by eight actuators.

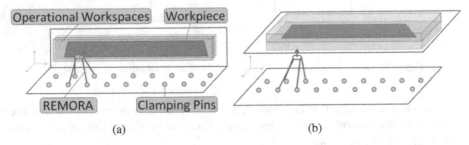

(a)   (b)

Fig. 9. (a) Sideward Workspace; (b) Upward Workspace

For different applications, the location of the workspaces of the robot can be different according to the way the tool is installed on the PP. Fig. 9 shows two kinds of workshop settings. The semi-transparent zone can be considered as the operational workspace, where the tool of the robot is in contact with the workpiece during operation. In order to avoid collisions with workpieces, the robot should not enter such a zone during the locomotion phase. Thanks to the presence of the kinematic redundancy, it becomes possible to take such kind of constraints into account. Fig. 10 shows a simulation example where the robot is achieving a series of steps while the

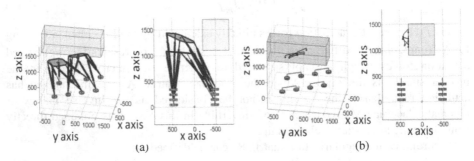

(a)   (b)

Fig. 10. (a) Key poses configurations; (b) Trajectories of the center of the PP

PP avoids entering the predefined machine zone. Fig. 10-b shows the trajectory of the center of the PP during the locomotion. One can notice that these trajectories stay outside of the predefined 'forbidden' zone. On the other hand, these trajectories keep close to the boundary in order to avoid reaching the limits of actuators strokes. The optimization algorithm which generates such trajectories as well as the management of the kinematic redundancy in REMORA has been detailed in [7].

4   Prototype: REMORA

To demonstrate the feasibility of designing a walking robot based on parallel robots with the help of lockers and clampers, a prototype has been built (Fig. 11). It is worth mentioning that most components used on the prototype are standard parts that can be found in catalogs. The principal characteristics of the robot can be summarized as follows:

- Eight electric actuators are distributed on the four limbs of the robot. Spur gears are used for transmission between the ball screws and the motors (Fig. 12-a).
- Three electric clutches (Fig. 12-b) are equipped on each limb, so there are 12 clutches in total, plus one brake on each motor for security. Because of difference required locking forces, the clutches near the extremity of the limbs have friction disks, while those on the PP side are with dented disks.
- A pneumatically actuated clamping device is used at the extremity of each limb (Fig. 12-c). In industry, these clamping devices are usually used for holding parts to be machined. In terms of clamping force, repeatability and misalignment tolerance, they are the ideal choice for our application.

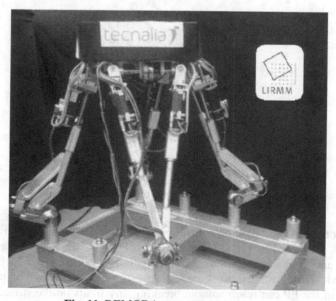

Fig. 11. REMORA prototype

(a) Motors and screws   (b) Clutches (Lockers)   (c) Clampers

Fig. 12. Key components of REMORA

- The control system is based on xPCTarget which is a real time module in Simulink. An advantage of this system is that users with Simulink programming skills can execute Simulink models on a target computer for real-time testing applications.
- The stroke of the actuated prismatic joints is between 830 mm and 1309 mm.
- The distance between two opposite limbs on the platform is 558 mm and the distance between the two connection points of each limb to the platform is 359 mm.

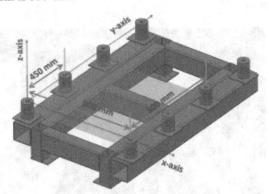

Fig. 13. Supporting **base and world frame**

The prototype is installed on a metal base. The world frame is arbitrarily fixed on the base as illustrated in Fig. 13. As shown in Fig. 14, the supporting pattern changes before and after one step, thus the corresponding workspaces of REMORA are also changed. The illustrated workspaces are estimated as the set of positions that the center of the PP can reach without breaking any limits on the strokes of the actuated prismatic joints and the limits of the passive revolute joints.

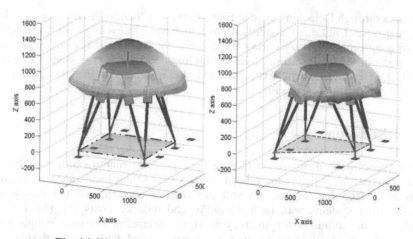

Fig. 14. Workspace variation under two supporting patterns

For realizing one step walking, the PP of the robot goes through five key poses: ($X_0 \rightarrow$ Initial pose; $X_1 \rightarrow$ Detaching pose; $X_1' \rightarrow$ Extraction pose; $X_2' \rightarrow$ Inserting pose; $X_2 \rightarrow$ Attaching pose).

Various tests have been achieved on the prototype. Fig. 15 shows the variations in the actuated prismatic joints plotted during one step. From t_0 to t_1, it is the reconfiguration phase where the robot works in PP mode, and the PP moves from the initial pose to the detaching pose. The length of every branch is changed. From t_1 to t_4, it is the limb swinging phase further divided into Extraction, (Free) Swinging and Insertion phase. The robot works in BE mode during those phases, and the PP of the

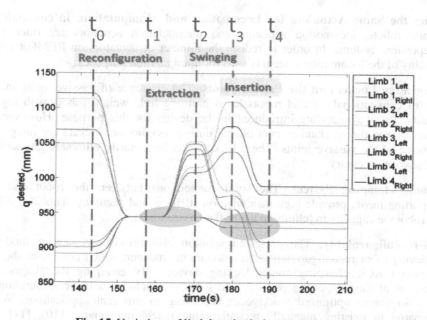

Fig. 15. Variations of limb lengths during a tested step

robot goes through Detaching pose to Extraction pose, then to Inserting pose, at last to Attaching pose. Unlike the standing limbs which vary in every phase, the lengths of the swinging limb only vary during (Free) swinging phase and remain steady during Extraction phase and Insertion phase (as highlighted by the two shadows in ellipse shape). This is because the extraction and the insertion of the clamping pin are achieved only by moving the PP.

5   Discussion

Mobile robots have difficulty in finding their place in manufacturing applications. One of the reasons is that many related issues such as their localization in a complex environment and locomotion are still not mature enough, while industry needs manufacturing systems with high reliability and reduced costs [1], [5]. Designing mobile robotic solutions for industry is about balancing between simplicity and functionality [8]. Such consideration has been applied both in the design of the structure and in the function modes of REMORA.

5.1   Novel Features in Mechanical Structure

Aiming to propose industry-oriented mobile robots, we have combined several techniques for sharing the advantages of parallel robots and legged robots.

Sharing Actuators for Positioning Each Leg. In conventional legged robots, five actuators are needed per leg to align its extremity to a point with the required orientation [9]. Instead of actuating each leg independently, we suggest moving the body of the robot to contribute positioning the extremity of the swinging leg.

Using the Same Actuators for Locomotion and Manipulation. In conventional mobile robots, locomotion actuators and manipulation actuators are often two independent systems. In order to reduce the number of actuators on REMORA, the mobility of the locomotion system is used for manipulation purposes.

Integrating Lockers on the Passive Joints. The existence of passive joints in the legs of conventional parallel robots helps building light-weight robots with higher rigidity. Passive joints are introduced in the design for this purpose. However, in order to keep the mechanism controllable during locomotion, lockers are integrated on some of the passive joints. These lockers can temporarily eliminate the passive DoFs when necessary.

Strong Clamping System. The solid connections between the robot and the supporting media provide high manipulation stiffness and accuracy. This also offers the robot the capacity to 'climb' in a non-flat environment.

Self-reconfigurability. The presented solution also provides a new method for achieving robot reconfiguration in an autonomous manner, which comprises the use of closed KCs, clamping and/or locking devices. By changing the shapes and locations of the supporting patterns, the robot can achieve self-reconfiguration in order to obtain optimized workspaces according to different applications. When compared to existing manually reconfigurable machining centers [10], [11], the suggested family of robots requires much less machine downtime by achieving reconfiguration in an autonomous way. The reconfigurability can also be interpreted from the robotics modeling point of view: the fact that robot respects same geometry constraints allows using the same mathematic models to describe the robot with different supporting patterns.

5.2   Potential Applications of REMORA

The presented robot illustrates new possibilities for the automation of some manufacturing processes in large workspaces. In the solution developed here, the fact that the posture of each supporting patterns can be measured after installation helps avoid localization issues. The precision of the prototype is partially ensured by measuring directly the position of the preinstalled clamping pins. Such measurements are carried out using common equipment such as laser trackers. As a result, when compared to the approach of high stiffness rail-based heavy machining centers, this method appears attractive to the industry in terms of costs, agility and flexibility.

In addition, allowing extending workspace of parallel robots to an unlimited large workspace can be interesting for many manufacturing applications, particularly when required charges and/or dynamics performances are difficult to fulfill with the conventional mobile arm-based solutions.

5.3   Future Work

As mentioned previously, the proposed robots can be reconfigured in order to customize the robot's performance according to application requirements. The arrangements of the supporting patterns will influence the force performance as well as the size of the local workspace during the machining mode. An advanced algorithm which is able to choose the optimal supporting patterns with respect to the desired performance should be developed in order to exploit the potential of the self-reconfigurability of the robot.

On the other hand, constraints and required functionalities might vary considerably according to applications. To reply the specific constraints of various applications, new structures can be inspired from the summarized design principles.

6   Summary

The work reported in this paper is related to a kind of mobile manipulators which is able to achieve locomotion tasks and to accomplish operations once it is deployed in its working location; in such way, some manufacturing operations can be automated in very large workspace. Several aspects such as working modes, working scenario as well as redundancy are addressed briefly. A prototype has been built and tested to illustrate the feasibility of the concept. Future work on the optimal design of the arrangement of supporting patterns as well as new structures design are expected for applications with specific requirements.

References

1. Stillstrom, C., Jackson, M.: The Concept of Mobile Manufacturing. Journal of Manufacturing Systems 26, 188–193 (2007)
2. Spenneberg, D., McCullough, K., Kirchner, F.: Stability of Walking in a Multilegged Robot Suffering Leg Loss. In: IEEE International Conference on Robotics and Automation, Barcelona, vol. 3, pp. 2159–2164 (2004)
3. Matsumoto, O., Kajita, S., Saigo, M., Tani, K.: Dynamic Trajectory Control of Passing Over Stairs by a Biped Type Leg-Wheeled Robot with Nominal Reference of Static Gait. In: IEEE/RSJ International Conference on Intelligent Robots and Systems, Victoria, vol. 1, pp. 406–412 (1998)
4. Buehler, M., Playter, R., Raibert, M.: Robots Step Outside. In: International Symposium on Adaptive Motion in Animals and Machines, Ilmenau (2005)
5. Armada, M., Prieto, M., Akinfiev, T., Fernandez, R., Gonzalez, P., Garcia, E., Montes, H., Nabulsi, S., Ponticelli, R., Sarria, J., Estremera, J., Ros, S., Grieco, J., Fernandez, G.: On The Design And Development Of Climbing And Walking Robots For The Maritime Industries. Journal of Maritime Research 2(1), 9–32 (2005)
6. Corbel, D., Gouttefarde, M., Pierrot, F.: Towards 100G with PKM. Is Actuation Redundancy A Good Solution for Pick-And-Place? In: IEEE International Conference on Robotics and Automation, Anchorage, pp. 4675–4682 (2010)
7. Yang, H., Krut, S., Baradat, C., Pierrot, F.: Locomotion Approach of REMORA: A Reconfigurable Mobile Robot for Manufacturing Applications. In: IEEE/RSJ International Conference on Intelligent Robots and Systems, San Francisco, pp. 5067–5072 (2011)
8. Yoneda, K., Ota, Y.: Non-Bio-Mimetic Walkers. International Journal of Robotics Research 22(3-4), 241–249 (2003)
9. Balaguer, C., Gimenez, A., Jardon, A.: Climbing Robots Mobility for Inspection and Maintenance of 3D Complex Environments. Autonomous Robots 18(2), 157–169 (2005)
10. Xi, F., Li, Y.W., Wang, H.B.: A Module-Based Method for Design and Analysis of Reconfigurable Parallel Robots. In: International Conference on Mechatronics and Automation, Bangkok, pp. 627–632 (2010)
11. Bi, Z.: Development and Control of a 5-Axis Reconfigurable Machine Tool. Journal of Robotics 2011, arti. 583072, 9 (2011)

Analysis and Optimal Design of a Spherical Parallel Manipulator with Three Rotational Degrees of Freedom

Bing Li[1], Shucan Chen[1], and Dan Zhang[2]

[1] Shenzhen Graduate School
Harbin Institute of Technology
Shenzhen, 518055, China
libing.sgs@hit.edu.cn
[2] Faculty of Engineering and Applied Science
University of Ontario Institute of Technology
Oshawa, Ontario, 2000, Canada
dan.zhang@uoit.ca

Abstract. Spherical parallel manipulators with three rotational degrees of freedom have been widely researched for a long time, and many mechanisms have been proposed and designed for some specific tasks. In this paper, the concept design of a spherical parallel manipulator with three rotational degrees of freedom is presented. The optimal design of this type of parallel mechanism with a 3SPU-1S kinematic chain is detailed which can act as a haptic device, and the purpose of the optimal design is to find a set of parameters that achieve a relatively good performance in terms of three important indexes, that is, to define the tradeoff among the workspace capabilities, dexterity and stiffness. First, the inverse kinematic equations and Jacobian matrix of the spherical parallel manipulator are formulated, which is the necessary for subsequent analysis. Then, three important performance is analyzed. Finally, the dimensional synthesis based on a compound performance index is introduced and simulation results are obtained.

Keywords: Spherical parallel manipulator, performance indexes, optimal design.

1 Introduction

Spherical parallel manipulators [1] are widely used in many applications that require the mobile platform to rotate about a fixed point arbitrary, and already give rise to interesting applications by orienting machine tools beds and workpieces, surgical tools and shoulder mechanisms, etc. As early as in 1928, Gwinnett [2] put forward a kind of three degrees of freedom spherical parallel mechanism, which is used as a platform for a movie theater. In 1994, Gosselin [3] discussed the 3-RRR parallel manipulator named Agile eye devices, which is applied to the camera positioning.

P. Neto and A.P. Moreira (Eds.): WRSM 2013, CCIS 371, pp. 71–81, 2013.

Later on, Karouia and Hervé studied the structural synthesis of asymmetrical non-over constrained 3-DOF spherical parallel mechanisms [4]. However, the applications of these spherical parallel manipulators are limited due to their stiffness. Then, a 3-legged UPS spherical parallel manipulator with the S limb highly increase its stiffness was studied, and in 2002, the optimization of the mechanism in terms of workspace also carried out [5].

Haptic devices are useful and have several potential applications such as, fine compliant assembly, VR environment simulation, and surgery, especially in hazardous or hostile areas. Recently the need for haptic devices with high-performance is raising due to the increasing usage of virtual environments capable to perform dexterous manipulation tasks. Parallel mechanisms have many advantages over serial mechanisms in terms of stiffness and precision and are therefore especially appropriate for high-performance haptic devices. Moreover, in some cases, for example, a haptic interface that best emulates the human wrist, the 3-DOF spherical parallel manipulators with large workspace and high stiffness are required [6].

In this work, we focus on a 3SPU-1S Spherical parallel manipulator, herein, S, P, U denote the spherical joint, prismatic joint and universal joint, respectively, P means the actuated joint. This kind of mechanism can be used as a haptic device with only three rotational degrees of freedom to provide users with feedback information on the motion and/or force that he or she generates. We try to find the geometric parameters of the mechanism that guarantee a relatively good performance. The challenge of the optimal design is not only consider just one performance index, but the other performance indexes should also be considered during the design of the parallel robot. The most known and used indexes includes the global condition index and the global stiffness index [7].

The remainder of this paper is organized as follows: the inverse kinematics of the 3-DOF is presented, and the Jacobian matrix is derived by the use of screw theory in Section 2. Then in Section 3, the reachable workspace is obtained and the three performance indexes are proposed. Next, the optimal design in terms of a novel compound index is introduced and simulation results are presented in Section 4. Finally, the conclusions are stated in Section 5.

2   The 3 SPU-1S Spherical Parallel Manipulator

Fig. 1(a) shows the CAD model and the schematic diagram of the 3SPU-1S spherical parallel manipulator. The four-legged mechanism is controlled by three SPU type legs with the prismatic pair actuated, while the fourth leg with a passive S joint is in the middle of the parallel platform, thereby giving the mobile platform with three rotational motions in space. Moreover, we can fully described the geometry of the parallel mechanism by the geometry parameters for convenient.

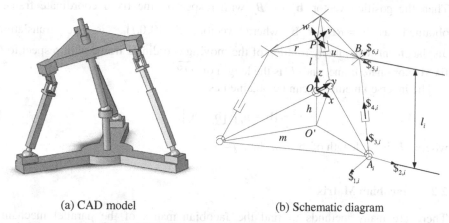

(a) CAD model   (b) Schematic diagram

Fig. 1. 3SPU-1S spherical parallel manipulator

2.1   Inverse Kinematics

As we can see from Fig. 1(b), the origin O of the fixed coordinate frame $O-xyz$ is located at the center of spherical joint of the middle leg with the x axis parallel to the direction of $\overline{O'A_1}$, where O' is the centroid of $\triangle A_1A_2A_3$, and $|O'O|=h$. Similarly, the origin P of the moving coordinate frame $P-uvw$ is attached to the centroid of $\triangle B_1B_2B_3$ with the u axis pointing in the direction of $\overline{PB_1}$. We assume that $|OA_1|=|OA_2|=|OA_3|=m$ and $|PB_1|=|PB_2|=|PB_3|=r$, where points A_i and B_i, $i=1,2,3$ are at the center of the universal joints and spherical joints of each leg respectively.

Let $\mathbf{a}_i$ represent the position vector of points A_i with respect to the fixed coordinate frame, it is expressed as

$$\mathbf{a}_1=\left[m,0,-h\right]^T, \quad \mathbf{a}_2=\left[-\frac{1}{2}m,\frac{\sqrt{3}}{2}m,-h\right]^T, \quad \mathbf{a}_3=\left[-\frac{1}{2}m,-\frac{\sqrt{3}}{2}m,-h\right]^T. \quad (1)$$

Similarly, $^P\mathbf{b}_i$ represents the position vectors of points B_i with respect to the moving coordinate frame can be expressed as

$$^P\mathbf{b}_1=\left[r,0,0\right]^T, \quad ^P\mathbf{b}_2=\left[-\frac{1}{2}r,\frac{\sqrt{3}}{2}r,0\right]^T, \quad ^P\mathbf{b}_3=\left[-\frac{1}{2}r,-\frac{\sqrt{3}}{2}r,0\right]^T. \quad (2)$$

According to the Tilt-and-Torsion ($T\&T$) angle method [8], s represents sine and c denotes cosine, the rotation matrix of the $T\&T$ angles can be written directly as

$$R(\varphi,\theta,\psi)=\begin{bmatrix} c\varphi c\theta c(\psi-\varphi)-s\varphi s(\psi-\varphi) & -c\varphi c\theta s(\psi-\varphi)-s\varphi c(\psi-\varphi) & c\varphi s\theta \\ s\varphi c\theta c(\psi-\varphi)+c\varphi s(\psi-\varphi) & -s\varphi c\theta s(\psi-\varphi)+c\varphi c(\psi-\varphi) & s\varphi s\theta \\ -s\theta c(\psi-\varphi) & s\theta s(\psi-\varphi) & c\theta \end{bmatrix}. \quad (3)$$

Then the position vector $\mathbf{b}_i$ of B_i with respect to the fixed coordinate frame is obtained as $\mathbf{b}_i = R\mathbf{P} + R^P\mathbf{b}_i$ where vector $\mathbf{P} = [0,0,l]^T$ is the translational displacement without any rotation of the moving coordinate frame with respect to the fixed coordinate frame, and l is the length of $\overline{OP}$

The inverse kinematics can be obtained as

$$l_i^2 = [\mathbf{b}_i - \mathbf{a}_i]^T [\mathbf{b}_i - \mathbf{a}_i]. \tag{4}$$

where, l_i is the length of the ith leg, $i = 1,2,3$.

2.2   Jacobian Matrix

There are many methods to find the Jacobian matrix of the parallel mechanism including differentiating the inverse kinematic equation. In this case, we use the screw theory approach and the concept of reciprocal screws [9].

We define an instantaneous reference frame $C - x'y'z'$ (which is not shown in the diagram) with its origin C attached to the point O and all the axes parallel to the axes of the fixed frame homologous. Thus we express all the infinitesimal twist of the mechanism with respect to this instantaneous reference frame. Let $s_{j,i}$ denote the unit vector along the jth joint axis of the ith leg, and $\hat{\$}_{j,i}$ represent the jth twist associated to the jth joint of the ith leg, then for $i = 1,2,3$ we have

$$\begin{cases} \hat{\$}_{1,i} = \left[\dfrac{s_{1,i}}{\mathbf{a}_i \times s_{1,i}} \right]; \hat{\$}_{2,i} = \left[\dfrac{s_{2,i}}{\mathbf{a}_i \times s_{2,i}} \right]; \hat{\$}_{3,i} = \left[\dfrac{s_{3,i}}{\mathbf{a}_i \times s_{3,i}} \right] \\[2mm] \hat{\$}_{4,i} = \left[\dfrac{0}{s_{4,i}} \right]; \hat{\$}_{5,i} = \left[\dfrac{s_{5,i}}{\mathbf{b}_i \times s_{5,i}} \right]; \hat{\$}_{6,i} = \left[\dfrac{s_{6,i}}{\mathbf{b}_i \times s_{6,i}} \right] \end{cases} \tag{5}$$

According to [8], each leg is considered as an open-loop chain and the instantaneous twists of the end effector in terms of the joint screws can be written as

$$\$_C = \begin{bmatrix} \mathbf{w}_C \\ \mathbf{v}_C \end{bmatrix} = \sum_{j=1}^{6} \dot{q}_{j,i} \hat{\$}_{j,i} . \tag{6}$$

where $\mathbf{w}_C = [w_x, w_y, w_z]^T$ denotes the angular velocity, $\mathbf{v}_C = [v_{Cx}, v_{Cy}, v_{Cz}]^T$ denotes the linear velocity, and $\dot{q}_{j,i}$ denotes the intensity. It is noticed that there is no translation for the end effector due to the passive spherical joint of the middle leg, thus $\mathbf{v}_C = [0,0,0]^T$.

According to the reciprocal screw's property, as the axes of the spherical and universal joints in each leg intersect the line passing through points A_i and B_i, a reciprocal screw is given as

$$\hat{\$}_{4,i} = \begin{bmatrix} \mathbf{s}_{4,i} \\ \mathbf{b}_i \times \mathbf{s}_{4,i} \end{bmatrix}. \tag{7}$$

herein, $\mathbf{s}_{4,i} = (\mathbf{b}_i - \mathbf{a}_i)/\parallel \mathbf{b}_i - \mathbf{a}_i \parallel$, then the reciprocal product is obtained as $\hat{\$}_{4,i}^T o\$_C = \dot{l}_i$ For legs $i = 1, 2, 3$, we have the velocity equation

$$\mathbf{J}_x \mathbf{w}_C = \mathbf{J}_q \dot{\mathbf{q}}. \tag{8}$$

where $\mathbf{J}_x = \begin{bmatrix} (\mathbf{b}_1 \times \mathbf{S}_{4,i})^T \\ (\mathbf{b}_2 \times \mathbf{S}_{4,i})^T \\ (\mathbf{b}_3 \times \mathbf{S}_{4,i})^T \end{bmatrix}$, $\mathbf{J}_q = \mathbf{I}_{3 \times 3}$ (3×3 identity matrix), $\dot{\mathbf{q}} = \begin{bmatrix} \dot{l}_1, \dot{l}_2, \dot{l}_3 \end{bmatrix}^T$.

Therefore, the velocity equation of the mechanism is expressed as $\mathbf{J}\mathbf{w}_C = \dot{\mathbf{q}}$, where the overall Jacobian matrix is $\mathbf{J} = \mathbf{J}_q^{-1} \mathbf{J}_x$.

3   Performance Analysis

3.1   Workspace Analysis

The workspace of the mechanism is an important criterion for analysis.

For the 3SPU-1S spherical parallel manipulator, it is more convenient to represent the reachable workspace in terms of $T \& T$ angles in a cylindrical coordinate system, where φ and θ are exactly the polar coordinates and ψ is the z- coordinate [7]. Here $\varphi \in [0, 2\pi)$, $\theta \in [0, \pi)$, and $\psi \in [0, \pi)$.

In this work, the constraint conditions are as follows:

Length limit of the legs: the leg length constraint of each leg is written as $l_{min} \le l_i \le l_{max}$, where l_i can be obtained from the inverse kinematic equation (4), while l_{max} and l_{min} denote the maximum and minimum strokes of the prismatic joint respectively.

Range of motion of the spherical and universal joints: for $i = 1, 2, 3$ we define θ_{si} and θ_{ui} as the angles of the spherical and universal joints respectively, while $\theta_{s\,max}$ and $\theta_{u\,max}$ represent the maximum rotational angles of the spherical and universal joints respectively, $\theta_{si} \le \theta_{s\,max}$ and $\theta_{ui} \le \theta_{u\,max}$ are defined as the constraints.

Leg interference: we considered all the legs as cylindrical limbs, then the constraints are defined as $D_{oi} \ge (D_i + D_o)/2$ and $D_{pi} \ge (D_i + D_p)/2$,where D_i, D_o and D_p represent the diameters of the limb $A_i B_i$, $O'O$ and OP, respectively, D_{oi} denotes the shortest distance between $A_i B_i$ and $O'O$, D_{pi} denotes the shortest distance between $A_i B_i$ and OP.

Table 1. Dimensional parameters of the 3-DOF spherical parallel manipulator (unit: mm)

Parameter	m	r	h	l
Value	500	400	300	300

Table 2. Constraint variables of the 3-DOF spherical parallel manipulator (unit: rad or mm)

Variable	l_{min}	l_{max}	$\theta_{s\,max} = \theta_{u\,max}$	$D_i = D_o = D_p$
Value	400	750	$\pi/4$	30

The dimensional parameters and constraint variables are shown in Table 1 and Table 2, respectively, and the reachable workspace of the 3-DOF spherical parallel manipulator is demonstrated in Fig.2.

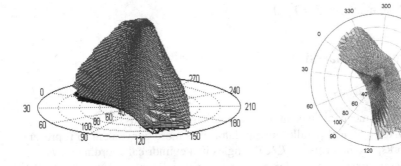

(a) Reachable workspace

(b) Vertical view of the reachable workspace.

Fig. 2. Reachable workspace

The workspace index of the mechanism is defined as one of the important criterions for evaluating the performance of the mechanism. It can be achieved by discretizing the feasible workspace into nodes n_f , and the total workspace into nodes n_t . In this case, we not only consider the physical limitations and conflict constrains, but also take the singular points and the singular approaching as obstacles when finding the feasible workspace. Hence the workspace index is obtained as the ratio of the two parameters

$$w = \frac{n_f}{n_t}. \tag{9}$$

It is obviously that the value of w is in the interval [0, 1].

3.2   Global Performance and Stiffness Analysis

Since the condition number is a local indication for the dexterity of a robot, it can be written as

$$k_J = \|J\| \|J^{-1}\| .$$ (10)

where $\|.\|$ denotes the norm of Frobenius, in order to evaluate the dexterity of a robot over a given workspace, the global condition index is introduced by Gosselin, as a performance index to evaluate the dexterity and isotropy with respect to the whole workspace[5]. The global condition index (GCI) is the reciprocal of the condition number of the Jacobian matrix integrated over the volume of the workspace and divided by the volume of the workspace. It is defined as

$$\eta = \frac{\int_w \frac{1}{k_J} dw}{\int_w dw} .$$ (11)

Nevertheless, it will be a big work to get the exact solution as the integral term will take a lot of time, then we define a discrete formulation instead. As we have discretized the feasible workspace into nodes n_f, then the modification of equation (11) is given below

$$\eta = \frac{\sum_{j=1}^{n_f} \frac{1}{k_J}}{n_f} .$$ (12)

where η takes values over the interval (0, 1].

Stiffness is also one of the important criterions of the parallel mechanisms, and it has a great impact on the dynamic accuracy and affects control performance. The stiffness matrix is configuration dependent and can be presented as $K = kJ^T J$. In order to perform an optimal design with the stiffness considered in the design stage, it is desired to analyze the 3-DOF spherical parallel manipulator's stiffness characteristics over the workspace. Here, we assume that the spring constant factor $k = 1$, then the stiffness matrix is expressed as $C = J^T J$ instead.

Similarly the global stiffness index (GSI) is the reciprocal of the condition number of the stiffness matrix integrated over the volume of the workspace and divided by the volume of the workspace. It is defined as

$$\xi = \frac{\int_w \frac{1}{k_C} dw}{\int_w dw} .$$ (13)

It is also be replaced as

$$\xi = \frac{\sum_{j=1}^{n_f} \frac{1}{k_C}}{n_f} \ .$$

(14)

where $k_C = \|C\|\|C^{-1}\|$, and ξ takes values over the interval (0, 1].

4   Optimal Design

4.1   Design Variables and Constraints

The moving platform depends on four length of the elements m, r, h and l according to the mechanism's geometric characteristic. For the sake of having a dimensionless analysis, three parameters are introduced as

$$f_1 = \frac{r}{m}, \quad f_2 = \frac{l}{m}, \quad f_3 = \frac{h}{m} \ .$$

(15)

Since we set the fixed frame coordinate at the center of spherical joint of the passive leg, the increase of the parameter f_3, means the elongation of the actuated legs, in this case, we set $f_3 = 0.5$ as a constant number. Therefore, the dimensional synthesis is boiled down to the influence of the two scale parameters f_1 and f_2. In consideration of the sufficient workspace and stiffness, the allowable values of them is defined as $f_1 \in [0.3, 1.5]$ and $f_2 \in [0.3, 1.5]$

Furthermore, the region of the reachable workspace can be obtain by the method of Section 3, while the $T \& T$ angles are constrained as $\varphi \in [0, 2\pi)$, $\theta \in [0, 4\pi/9]$, and $\psi \in [0, \pi/3]$. Besides, the angle scope of the spherical and universal joints θ_{si} and θ_{ui} should both be constrained within the angle $\pi/4$.

4.2   Compound Index

The optimal design of the 3S<u>P</u>U-1S spherical parallel manipulator is based on the three indexes: workspace index, global condition index and global stiffness index.

As we can see from equation (9), (12) and (14), the global condition index and the global stiffness index as the most known global performance indexes (GPI) both have a relationship with the workspace, and the performance of the spherical parallel manipulator is evaluated over the workspace which is discretized into nodes for the sake of simplicity of computation. With the purpose of evaluating the performance in terms of the workspace capabilities, dexterity and stiffness, the compound index is introduced as follows

$$C = \alpha w (\sum_i^n \beta_i GPI_i) \ .$$

(16)

where α and β_i are weights, and for this 3S$\underline{P}$U-1S spherical parallel manipulator, It can be defined as

$$C = w(\eta + \xi). \tag{17}$$

Then our goal is to maximize C in the balance of w, η and ξ.

4.3   Simulation Results

Fig. 3 illustrates the workspace index, global condition index and global stiffness index distribution in terms of the scale parameters f_1 and f_2, respectively. The optimal is presented in Table 3.

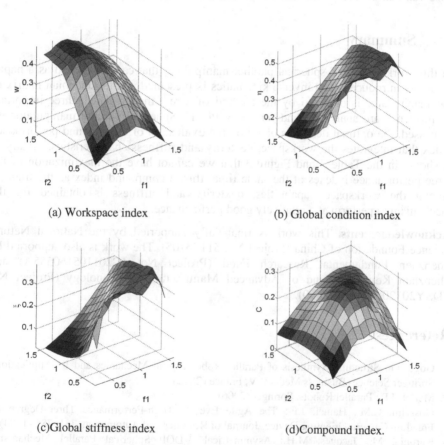

(a) Workspace index   (b) Global condition index

(c)Global stiffness index   (d)Compound index.

Fig. 3. The distribution of the index with regard to f1 and f2

Table 3. Performance indexes results

Index	w	η	ξ	C	$[f_1, f_2]$
w_{max}	0.4979	0.2203	0.0636	0.1414	$[0.6, 1.4]$
η_{max}	0.1111	0.5701	0.3887	0.1065	$[1.3, 0.3]$
ξ_{max}	0.1111	0.5701	0.3887	0.1065	$[1.3, 0.3]$
C_{max}	0.2835	0.4651	0.2836	0.2835	$[0.9, 0.7]$

The maximum value of the compound index is $C_{max} = 0.2835$, it turns up at $f_1 = 0.9$, $f_2 = 0.7$. As we can see from Table 3 and Figure 3, it can be easily observed that the compound index is the tradeoff of the three performance.

5   Summary

In this work, a 3-DOF spherical parallel manipulator that can be applied as a haptic device is introduced. The inverse kinematics is presented and the Jacobian matrix of the manipulator is obtained by the method of screw theory. Then, three important performance are analyzed. And for the 3SPU-1S Spherical parallel manipulator we discussed, the optimal design is based on the evaluation of a compound performance index that integrates the workspace, dexterity and stiffness characteristics. Finally, it is shown in the Table 3 and Figure 3 that we cannot have the maximization of the three performance indexes at the same time, thus a compound index as the tradeoff among the workspace capabilities, dexterity and stiffness is obtained for the mechanism to perform at a relatively good performance.

Acknowledgements. This work is financially supported by the National Natural Science Foundation of China (Project No. 51175105). The work is also supported by Shenzhen Fundamental Research Fund (Project No. JC201105160555A) and Shenzhen Key Lab Fund of Advanced Manufacturing Technology (Project No. ZDSY20120613125132810).

References

1. Gogu, G.: Structural Synthesis of Parallel Robots, Solid Mechanics and Its Applications. Springer Science+Business Media B.V, France (2012)
2. Merlet, J.P.: Parallel Robots. Springer (2006)
3. Gosselin, C.M., Hamel, J.F.: The Agile Eye: A High-Performance Three-Degree of-Freedom Camera-Orienting Device. Journal of Robotics and Automation, 781–786 (1994)
4. Karouia, M., Jacques, M.H.: Asymmetrical 3-DOF Spherical Parallel Mechanisms. European Journal of Mechanics A/Solids 24, 47–57 (2005)
5. Badescu, M., Morman, J., Mavroidis, C.: Workspace Optimization of Orientational 3-legged UPS Parallel Platforms. In: Proceedings of DETC 2002 ASME 2002 Design Engineering Technical Conferences and Computers and Information in Engineering Conference Montreal, Canada (2002)

6. Birglen, L., Gosselin, C., Pouliot, N.: SHaDe, a New 3-DOF Haptic Device. IEEE Transactions on Robotics and Automation 18(2) (April 2002)
7. Liu, X.J., Jin, Z.L., Gao, F.: Optimum Design of 3-DOF Spherical Parallel Manipulators with Respect to The Conditioning and Stiffness Indices. Journal of Mechanism and Machine Theory 35, 1257–1267 (2000)
8. Bonev, I.A., Ryu, J.: Orientation workspace analysis of 6-DOF parallel manipulators. In: Proceedings of the 1999 ASME Design Engineering Technical Conferences, Las Vegas, Nevada, September 12-15 (1999)
9. Tsai, L.W.: Robot Analysis: The Mechanics of Serial and Parallel Manipulators. A Wiley-Interscience Publication, Canada (1999)

Stereoscopic Vision System for Human Gesture Tracking and Robot Programming by Demonstration

Marcos Ferreira, Luís Rocha, Paulo Costa, and A. Paulo Moreira

INESC-TEC (formerly INESC Porto)
Rua Dr. Roberto Frias, s/n, Campus FEUP
{marcos.a.ferreira, luis.f.rocha}@inescporto.pt
{paco, amoreira}@fe.up.pt

Abstract. This paper presents a framework for robot programming by demonstration using gesture. It is based on a luminous multi-LED marker which is captured by a pair of industrial cameras. Using stereoscopy the marker supplies a complete 6-DoF human gesture tracking output with both position and orientation. Tests show that the developed setup is industrial grade, being precise for many industrial applications and robust particularly to lighting conditions. Attaching the marker to an operator work tool provides an efficient way to track the human movements without further intrusion in the process. The resulting path is used to generate a program for an industrial manipulator ending the cycle in an human-robot skill transfer framework.

Keywords: Programming-by-demonstration, motion tracking, artificial vision, skill-transfer, robotics, industrial manipulators.

1 Introduction

Production lines tend to evolve into the concept of mass customization, i.e., working on small series with adapted and specialized procedures to each of them according to costumer specific needs. High versatility is mandatory in these systems and robotised cells demand additional efforts to be integrated in such systems: industrial manipulators still take a long time to reconfigure. Programming is extremely time consuming and usually require experienced and highly qualified workers. Overall this is not compatible with flexible setups neither with companies' budgets since both qualified programmers and reconfigurations downtime imply strong financial efforts.

Even though the former is a quite restrictive scenario, manipulators are still strongly desired at production lines due to a series of advantages over human work, e.g., the ability to work continuously, the high accuracy and repeatability, immunity to fatigue, distractions and hazardous environments.

To overcome this reality, this work presents a methodology for fast industrial robot programming via human demonstration by gesture. The main goal is to achieve a framework where a specialized human operator shows the robot how to do a concrete task with abstraction of the programming language and even completely avoiding the use of the teach pendant.

P. Neto and A.P. Moreira (Eds.): WRSM 2013, CCIS 371, pp. 82–90, 2013.

The focus of this paper is to describe the learning by demonstration frame-work: a pair of industrial cameras is used to retrieve 6D data of an human gesture through the use of stereoscopy. The artificial vision system captures images of a specific luminous marker that can accurately tell us information of position and orientation. The resulting set of points that describe the human path are automatically transformed in a robot program. The operator uses his natural abilities and skills to accomplish the demonstration process without needing further instruction on using new software packages, interfaces or tools.

1.1   Related Work

Robot programming from demonstration (PbD) has captured a lot of attention in the late years. In [1] we can find a survey on PbD which categorizes the most used styles/methodologies. According to those categories our work is best described as an imitation learning process. In this area, we can find: PbD by gesture tracking for recognition, PbD using voice commands, CAD models, force/torque control and, more generically, demonstration using artificial vision, image recognition and other vision based systems.

In the context of PbD through gestures, authors usually try to differentiate a set of gestures, recognize them and then have the robot perform movements according to each gesture. Such is the work of [2] and [3]. In [2], the author uses Hidden Markov Models to distinguish between gestures whilst, in [3], we see the use neural networks while maintaining the same base idea of using a wearable device to capture moves. These still fail to provide abstraction from the robot programming language since the gestures must be pre-programmed in the robot.

Using voice commands is another approach that has been tried in [4]. Again, the code for the robot's actions must be previously written and voice commands are used to choose from an existing set of movements.

Other contributions include the use joysticks and modified commercial tools people are used to handle every day as digital pens([5]). The work described on [6] and [7] refer to telemanipulation; while not being a clear way of programming by demonstration, it is still an intuitive way to interface and control robot manipulators. Most of these solutions can be seen applied to medical robotics, as in surgical interventions. In [8] and [9]we can find PbD using CAD. The described setup is very user friendly yet it still forces the operator to learn how to use the software which can be problematic with non-skilled operators in real industrial scenarios. Authors of [12] present a system architecture where the CAD based offline programming is complemented with an automatic object recognition: the programs are generated through intuitive CAD interface and the laser-scanning module identifies objects and uploads the corresponding machine code into the robot controller. The work described in [10] and [11] presents a new input device: a set of infrared markers that can be attached to any kind of tool in a concept similar to the one presented in this paper; a set of cameras do the tracking of the markers retrieving their position during a demonstration. Although it is a promising approach, infrared cameras are far more expensive than their RGB counterpart, and typically have lower resolutions.

Moreover, the robot controller is made using low-level interfaces with the motor controllers. This type of control in not available on o_-the-shelf industrial robots or it overrides their safety circuits. In our project we implement a translation mechanism that generates robot programs that any qualified user can navigate and edit if necessary.

2   Tracking Human Motion

2.1   Background - *Sincrovision*

The *sincrovision* concept was developed and patented in the University of Porto - Faculty of Engineering. It implements a system of 3D acquisition based on stereoscopic vision synchronised with high intensity luminous markers [13]. The key idea is to turn on the light emitters as soon as the cameras start acquiring image and turn them o_ after the camera exposure time has expired. Figure 1 (left) shows a timing diagram of the system.

The high intensity lights will be very bright on the images whilst the back-ground noisy data will have no time to be acquired by the camera. At the same time, blinking the markers for a short time makes it possible to stare at them; keeping them always on would cause eye damage. To ensure that unwanted light is not captured, the lenses aperture is reduced to minimal. This setup makes it possible to triangulate the markers positions in space in a robust way, independently of lighting conditions in the scene and ignoring most of the common noise sources in artificial vision applications. Figure 1 (right) shows a typical image captured by the cameras using this synchronous feature.

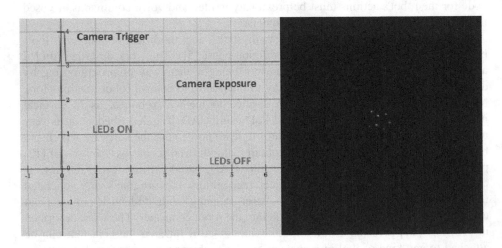

Fig. 1. Left:Camera and LEDs triggers: timing diagram; LEDs remain active while the camera is acquiring image. Right: A video frame showing a set of LEDs captured using the *sincrovision* concept; the lights in the room don't show up. Overall it is almost noise free.

The gesture tracking framework that has been developed is based on a luminous marker that implements the *sincrovision* concept. Taking advantage of the accurate measures of the stereo pair, we intend to capture every wrist move of an human operator and send it to a robotic manipulator.

2.2   A Marker for 6-DOF Movement Capture

The developed marker is based on 20 high intensity RGB-LEDs - Fig. 2.Whilst a single LED can provide the position data, a large group of markers also enables capturing the orientation with the added benefit that every single one can contribute for a better estimate of the position. Theoretically 3 LEDs were enough to retrieve 6D data though it would fail quite often due to occlusions. We used the geometric figure of an icosahedron as a base for the marker as it provides an interesting set of properties:

-    Placing the LEDs of the centre of each face covers full 360 degrees rotations in every axis; there are always at least 3 lights visible to compute orientation
-    The centre of the faces | or the vertexes of a dodecahedron (dual polygons) - can be described in spherical coordinates meaning that the LEDs are all positioned on a sphere shell. This property is useful for algorithmic purposes as will be described ahead.
-    Regarding construction issues, this polygon is a balanced (symmetric) object that easily attaches to an industrial tool (Fig. 2). Moreover, the empty space in its interior makes room for electronics, LEDs' supports and easy cabling.
-    The symmetry makes its centre rotation invariant which decouples the estimation of position and orientation.

Fig. 2. Left: A CAD model of the developed icosahedron-shaped marker. *Right:* The real icosahedron with all LEDs turned on. The stereoscopic setup can also be been in background. In this picture the marker was attacked to a painting spray gun.

- Distribution of the LEDs in a grid/net pattern. From the graph representation of a dodecahedron, one can colour the vertexes using only 5 distinct colours and obtain unique sets: a vertex and its 3 neighbours are unique throughout the whole marker (see Fig.3). This property is used to estimate orientation as described further ahead.

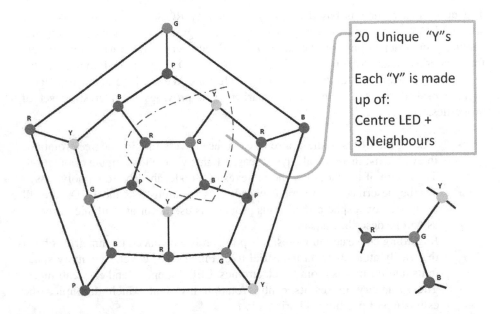

Fig. 3. Graph representation of a dodecahedron. The chosen colour pattern guarantees unique "Y"s. These colours are applied with the RGB LEDs. The vertexes positions are the same as the face centres of the icosahedron.

2.3   Estimating Translation and Orientation

One of the key characteristics of the marker is its symmetry and the fact that the LEDs occupy positions along a surface of a sphere. At any given time (image frame), and neglecting the colour of the LED, we compute the 3D positions in space of all visible lights using the stereo principles.

With that set of world coordinates we use constrained least squares estimation (cLSE) to determine both the position of the centre of the sphere and the length of its radius - Equation 1.

$$(x_i - x_c)^2 + (y_i - y_c)^2 + (z_i - z_c)^2 - r^2 = 0. \tag{1}$$

(Further development of this equation into the constrained least squares matrix form is fully described in [14]).

Even though the radius is known a priori, its estimation gives us indication of the quality of the LSE result. If it significantly differs from the well-known constructive measure, it indicates that some of the stereo points are bad. Iteratively dropping points and redoing the LSE we achieve a robust estimation of the sphere centre.

Concerning orientation, we use the coloured graph representation of the dodecahedron as seen in Fig. 3. Each RGB LED is driven into the required colour and we start o_ by creating a list with the 3D world coordinates of each LED and its neighbour's colours.

When processing the image frames, we look for a complete set of a LED and its 3 neighbours; then we match that set with an entry of our neighbour list: the matching is performed based on colour since each set is unique, as discussed above. After finding the correct entry on the list we know which LEDs are which, not only for those 4 but we also get to know all other LEDs in the image.

To compute the rotation from the stand-by position to the one captured in video, we use the Kabsch algorithm. On one hand we have the positions of all the LEDs in the video frame which makes our matrix Q; this data is collected from the stereo. On the other hand we have the 3D positions of those same LEDs in the stand-by pose (the vertexes of the dodecahedron); these vectors are arranged in matrix P so that Q and P are paired: the nth vector in Q is the new world position of nth vector in P (the stand-by position).

The Kabsch algorithm finds the rotation matrix that optimally describes (in a root-mean-squared error sense) the rotation from two paired of 3D point lists:

1. P and Q must have origin-centred vectors so the first step is subtracting both sets their respective centroid.
2. Compute the covariance matrix A defined as: $A = P^T Q$
3. Compute the optimal rotation matrix U:

$$A = VSWT \; ; \; d = \text{sign}(\det(WVT)) \; ; \; \rightarrow \; U = W \begin{pmatrix} 1 & 0 & 0 \\ 0 & 1 & 0 \\ 0 & 0 & d \end{pmatrix} V^T. \qquad (2)$$

(The auxiliary parameter d is used to insure we get a right handed coordinated system).

At this point we have the full characterization of the movement with both position and orientation.

2.4 Interface with Industrial Manipulator

The set of 6-D points is used to build a program for a robotic manipulator. We performed tests with a Motoman painting-specific robot; for that purpose, we implemented a module that creates the required set of movement instructions which make the robot mimic the captured human gesture - Fig. 4.

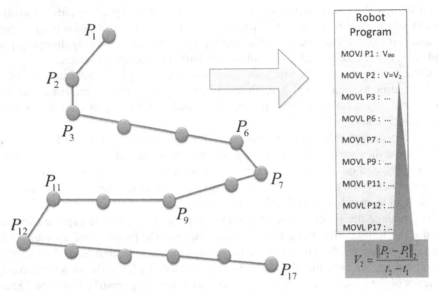

Fig. 4. Automatic generation of machine code from the motion positions captured by the stereo system. Each pose is associated with a time stamp given by the synchronous camera trigger signal thus providing a known linear velocity.

3   Results and Conclusions

3.1   Setup Description and System Precision

The required setup is not expensive as it can start with just a pair of cameras, the marker, LEDs and cheap electronics for timings. The precision of this framework largely depends of the number of cameras and their placement, along with resolution and marker size. Cameras placement have to be studied so they can both capture the entire workspace; it took us about half an hour to set the absolute positioning of these devices at our industrial demonstrator. Moreover, the *sincrovision* concept works on lens with fully closed aperture; in such configuration and working with 3ms of exposure time allows to eliminate all background noise; through several tests, there was no need to adjust these parameters - it behaved well both in laboratory tests and in a real industrial cell. Tests performed with 2 USB cameras (1024x768 @ 25fps), a marker with a radius of 50cm and a working volume of 1 cubic meter came up with the precision shown in table 1. This same set of components were tested in a real painting/coating application where the marker was coupled to an operator painting spray gun. The industrial robot executed the same movements and the precision was high enough so that the operators positively validated the final work of the robot. Note that path alone is not enough for a job description; in out test case as well as in other general application, additional data is required such as cell I/O's state: the gun

Table 1. Position and Orientation error of the spherical marker

	Mean Error	Max Error
Position	*3mm*	*5mm*
Orientation	1.5°	4°

trigger, conveyor and painting table movement and speed, all are synchronously recorded along with the pose data from video frames. When creating the robot program, all of these variables are inserted into the generated program and then they are set by the robot controller. This way, the robot mimics the full "picture" of the robotic cell - both human movement and external devices operation.

3.2   Discussion and Conclusion

The developed system provides an efficient and robust method for human gesture capture. IT allows a complete 6-DoF tracking and is immune to local lightning conditions; ceiling lamps and other types of luminous disturbance are eliminated. This overcomes one of the key conditionings in artificial vision applications.

Although the obtained path data can be used to diverse ends, this system provides a powerful framework for robot programming by demonstration, as shown by our tests with industrial manipulators.

The marker is easily attached to an operator working tool; in our test case we used a painting spray gun. No further changes are needed in the process so the human operator can demonstrate his skills without learning to work with software or any other tool other than his own. The pair of cameras takes little room to accommodate in an industrial cell so no deep and costly changes are needed compared to a normal robotized production cell.

Acknowledgements. Marcos Ferreira acklowledges FCT (Portuguese Foundation for Science and Technology) for his PhD grant | SFRH/BD/60221/2009. The work presented in this paper, being part of the Project PRODUTECH PTI (13851) - New Processes and Innovative Technologies for the Production Technologies Industry, has been partly funded by the Incentive System for Technology Research and Development in Companies (SI I&DT), under the Competitive Factors Thematic Operational Programme, of the Portuguese National Strategic Reference Framework, and EU's European Regional Development Fund. The authors also thank the FCT for supporting this work trough the project PTDC/EME-CRO/114595/2009 - High-Level programming for industrial robotic cells: capturing human body motion.

References

1. Argall, B., Chernova, S., Veloso, M., Browning, B.: A survey of robot learning from demonstration. Robot. Auton. Syst. 57(5), 469–483 (2009)
2. Ekvall, S., Kragic, D.: Grasp Recognition for Programming by Demonstration. In: IEEE International Conference on Robotics and Automation, Spain (2005)

3. Aleotti, J., Skoglund, A., Duckett, T.: Position Teaching of a Robot Arm by Demonstration with a Wearable Input Device. In: Proc. of Int. Conf. on Intelli. Manipul. and Grap., Genoa, Italy (2004)
4. Pires, J.N.: Experiments on commanding an industrial robot using the human voice. Indust. Robot: an Int. Journal 32(6), 505–511 (2005)
5. Pires, J.N., Godinho, T., Araujo, R.: Using Digital Pens to Program Welding Tasks. Ind. Robot, Emerald: Special Issue on Robotic Welding 6 (2007)
6. Monahan, E., Shimada, K.: Computer-aided navigation for arthroscopic hip surgery using encoder linkages for position tracking. International Journal of Medical Robotics and Computer Assisted Surgery (2006)
7. Rayman, R., Croome, K., Galbraith, N., McClure, R., Morady, R., Peterson, S., Smith, S., Subotic, V., Primak, S.: Long-distance robotic telesurgery: a feasibility study for care in remote environments. International Journal of Medical Robotics and Computer Assisted Surgery (2006)
8. Neto, P., Mendes, N., Arajo, R., Pires, J.N., Moreira, A.P.: High-level robot programming based on CAD: dealing with unpredictable environments. Industrial Robot, Emerald 39(3), 294–303 (2012)
9. Neto, P., Pires, J.N., Moreira, A.P.: CAD-based on-line robot programming. In: 4th IEEE Int. Conf. on Robotics, Automation and Mechatronics, Singapore, pp. 516–521 (2010)
10. Hein, B., Worn, H.: Intuitive and Model-based On-line Programming of Industrial Robots: New Input Devices. In: IEEE Int. Conf. on Intelligent Robots and Systems, USA (2009)
11. Hein, B., Worn, H., Hensel, M.: Intuitive and Model-based On-line Programming of Industrial Robots: A modular On-line Programming Environment. In: IEEE Int. Conf. on Robotics and Automation, Pasadena, USA (2008)
12. Ferreira, M., Moreira, A.P., Neto, P.: A low-cost laser scanning solution for flexible robotic cells: spray coating. The Int. Journal of Advanced Manufacturing Technology 58, 1031–1041 (2012)
13. 3D Object Motion Tracking and Locating System by Means of Synchronised Light Emitters with a Stereoscopic Vision System, http://v3.espacenet.com/textdoc?DB=EPODOC&IDX=WO2010046759&F=0
14. Chang, L., Pollard, N.: Constrained Least-Squares Optimization for Robust Estimation of Center of Rotation. Journal of Biomechanics (2006)

Development of a Flexible Robotic Cell for Laser Cutting of 3D Foam and Preformed Fabric

Félix Vidal[1], Rodrigo González[1], Marcos Fontán[1], Paula Rico[1], and Diego Piñeiro[2]

[1] AIMEN Technology Center, R/ Relva 27A,
36410 O Porriño, Spain
{fvidal,rgonzalez,mfontan,prico}@aimen.es
[2] SELMARK, PTL Valadares R/ C Nave 11,
36315 Vigo, Spain
diego.pineiro@selmark.es

Abstract. Laser technology is available nowadays for cutting of both textile and foam yielding high production levels and short setup times between cutting runs. However, the characteristics of the European Textiles and Clothing industry, that forces the design and manufacture of several Ready-to-wear collections each season, reduces the benefit of the laser technology due to the need of a continuous parameterization and reprogramming of the different clothing patterns. In this work, a flexible robotic laser cell for 3D cutting is presented, demonstrating that the production level and flexibility obtained can be advantageously used to improve the competitiveness of the European Textiles industry.

Keywords: Laser cutting technology, robotics technology, CAD-based programming, small-batch manufacturing, European Textile and Clothing Industry.

1 Introduction

1.1 Motivation

The European Textiles and Clothing industry represents one of Europe's major industrial sectors with an annual turnover of €235bn and a total workforce of 2.9 million in 2007 [1]. Besides, this sector is notable for having a predominance of Small and Medium-Sized Enterprises (SMEs) requiring significant human resources with a high proportion of female workers.

Globalisation and ongoing liberalisation expose EU industry to ever more competition from a large number of low-labour cost countries (especially from Southeast Asia) due to the combination of low wage costs with high-quality textile equipment and know-how imported from industrialised countries. The EU industry strives to remain competitive by means of higher productivity, and through competitive strengths such as innovation, quality, creativity, design and fashion [2]. These competitive advantages are the result of a permanent process of restructuring and modernisation.

P. Neto and A.P. Moreira (Eds.): WRSM 2013, CCIS 371, pp. 91–100, 2013.
© Springer-Verlag Berlin Heidelberg 2013

In this context, research on new manufacturing technology (e.g. laser technology) will be an important catalyst for industrial innovation, increasing the production level of the European Textiles sector enabling new product designs [3] [4]. Thus, the introduction of laser cutting technology offers a more profitable alternative to conventional cutting technologies due to its good accuracy, flexibility, quality and productivity. As a result, there are in the market different cutting machines, based on CNC and laser technology, capable to make 2D cutting of fabrics.

However, for applications that use 3D foam and/or preformed fabric (e.g. lingerie, footwear), currently there is no suitable commercial solution based on laser technology, since 3D cutting of these materials involves a time-consuming reprogramming of both laser parameters and 3D cutting paths.

Thus, the deployment of current laser-based cutting systems is not cost-effective due to the characteristics of the European Textiles and Clothing industry to design and manufacture different patterns of different collections and sizes each season, like small-batch manufacturing. Besides, European Textiles sector has difficulty finding skilled workers capable of operating with complex positioning systems like robots. Therefore, new and more intuitive ways for people to interact with these systems are required to make the programming of the cutting paths easier and accessible [5-7]. These facts establish a technological and economical barrier to the implementation of laser technology in these 3D applications.

To enable laser technology to penetrate in this kind of demanding market, a novel system involving laser technology, robotic and software development for control and path planning has been developed and integrated to define a new concept of flexible cell.

The flexible robotic cell proposed consists of an industrial robotic cutting system that can be easily extended with a CAD-based programming system that allows to the system to be easily reconfigured only loading the drawing of the pattern of the specific foam and/or preformed fabric to be cut. Thus, this flexible cutting system will extend the capabilities of the European Textiles sector increasing its competitiveness against low-labour cost countries [8].

1.2   Innovative Character

Industrial robots in real manufacturing scenarios are being programmed mainly manually through the use of a teach pendant. This approach requires a skilled robot programmer and turns unproductive the robot during this task. Besides, for complex parts and/or small batch manufacturing, like 3D cutting of preformed fabrics or preformed foam, the process of programming may take a long time. Therefore, the use of a robot may multiply the programming time instead of reducing it.

An alternative approach studied since the 80's that aims to achieve an intuitive programming procedure is learning from demonstration (LfD). Under this concept, very different techniques have been proposed in the context of both industrial and service robots [9] [10]. However, these approaches have a probabilistic nature that might introduce ambiguities and lack of the accuracy achieved by human operator in manual industrial processes, which prevents their use for laser processing.

Alternatively, offline environments are also being used to avoid the need to stop the robot in the programming tasks, since programming may be done on a PC while the robot is working [11]. However, since offline commercial simulation packages are valid for a wide range of applications, this approach still requires skills in offline robot programming and does not necessarily reduce programming time. Besides, the use of commercial simulation packages does not avoid the use of CAD modeling software.

In order to improve the negative points in the programming systems previously mentioned, in the work presented in this paper, an alternative, accurate and fast-programming approach has been developed. This approach is based on the characteristics of European SMEs to design and model their products before carrying out the manufacturing process improving their innovation potential [12].

In this paper, a new CAD-based robot programming system allowing operators with basic skills in CAD and robotics to generate offline robot routines is presented. In this approach the robot processing routines are extracted directly from the CAD model of the part to be processed. The extracted data are used to automatically generate the 3D robot paths for laser cutting processes. This CAD-based robot programming system has some relative advantages:

- Simplicity of use: The operator only has to load the CAD model into the software interface. Besides, the construction of CAD-model is obtained directly from design area.
- Low-cost solution: This approach avoids the use of simulation and offline programming software (e.g. RobotStudio, KUKA.Sim PRO, etc.).
- Accurate: The robot follows the desired cutting paths defined in CAD environment.

Therefore the operator only needs to work with a CAD environment to draw the theoretical positions of the cutting path and send the file to the system controller. The main limitation of this approach is that the real scenario has to reproduce exactly CAD model of the part to be cut. This requires a complex fixture system, combining mechanical and vacuum fixtures, to avoid displacements of the part (i.e. 3D bra cup).

2   Experimental Procedure

The main objective of the work presented is the development of a flexible robotic cell that will be able to make laser cutting of 3D foam and preformed fabric. Besides, this robotic system will be focussed on the particular characteristics of the Textiles sector that forces the design of several Ready-to-wear collections each season, like small-batch manufacturing.

Currently, the cutting process is a laborious semi-automatic cutting technique, based on the use of die cutters and subject to operator performance, which affects to the accuracy and quality obtained (Fig. 1).

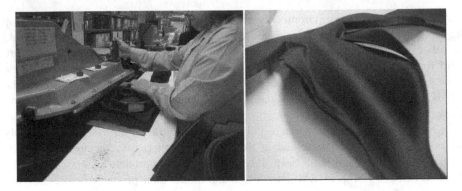

Fig. 1. Left: Current semi-automatic cutting process based on the use die cutters for 3D cutting of preformed foam. Right: Example of a cutting defect due to a wear of the die cutter.

Therefore, the development of a robotic cutting cell, based on laser technology, easily reconfigured and adapted for small-batch manufacturing, will increase the productivity and the competitiveness of the European Textiles sector, improving the working conditions of operators and the quality of the final part enabling new cutting applications.

2.1   Prototype Setup

The flexible robotic cell for laser cutting is made up of an IRB-4400 – 40/2,55 ABB robot equipped with a IRC5 controller, a CO_2 ROFIN OEM 10iX laser with a maximum power of 120W and 10.6μm wavelength, and a LASERMECH cutting head with a 4'' focal length and a 0.04'' diameter nozzle and 0.04'' tip stand-off (Fig. 2). Besides, laser cutting cell is equipped with an industrial computer running the developed software interface that extracts and generates robot routine from CAD file. This robot routine is sent to robot controller via TCP/IP connection.

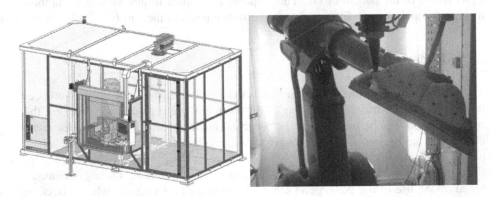

Fig. 2. Left: Final design of robotic cell for laser cutting of 3D preformed fabric and foam. Right: Example of laser cutting process.

The flexible robotic cell is configured to have the robot holding the part and move it while the laser cutting head is fixed in the structure of the robotic cell. This configuration simplifies the optical path of the laser system.

2.2   Intuitive Robot Programming Based on CAD

Small batch sizes present a particular challenge for robotic applications because of the need for programming each batch.

- Robot programming through the typical teaching method – usually, using a teach pendant— is a tedious and time-consuming task that requires technical expertise. Besides, this method requires the operator to manually define all poses (robot end-effector position and orientation) on the workpiece.
- Typical Off-line programming method also involves the employment of highly skilled labour in both robotics and mechanical processes.

The solution proposed in this work is to develop methodologies that help the operators to program a robot in an intuitive way, quickly, and with a high-level of abstraction from the robot language. In this system, the operator generates the robot routine simply drawing the robot cutting paths on the CAD environment (Fig. 3).

The developed solution was designed using AUTOCAD™ and DXF files [13] (Drawing Interchange Format) enabling data interoperability between AutoCAD and external programs.

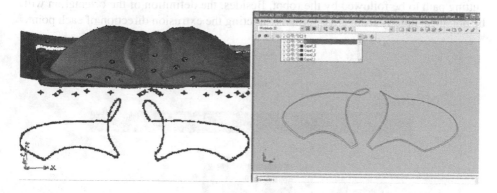

Fig. 3. Example of a CAD file used for the automatic generation of robot paths for the cutting of a preformed bra

After completing the design, CAD file is processed in order to generate a robot routine with the robot paths (including tool orientations and positions) and laser parameterization where laser cutting must be applied (Fig. 4). The definition of the orientation is made automatically normal to the surface of the preformed foam or fabric to be cut. This is because laser cutting process demands that the laser head has to be positioned perpendicular respect the part to be cut to avoid chamfers.

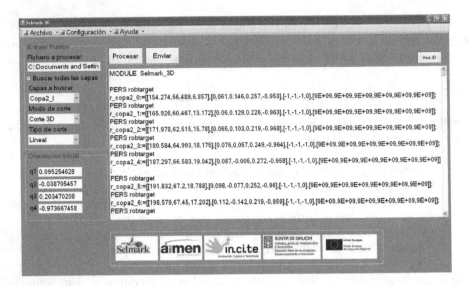

Fig. 4. Software application developed to extract, from CAD file, the robot routines automatically

2.3   Curve Fitting Algorithm

Cutting process begins with the point sequence extraction of the cutting path from the CAD model through DXF file format (Fig. 5). This point sequence will define the cutting path to be followed by the robot. Besides, the definition of the orientation will be obtained directly from DXF format extracting the extrusion direction of each point.

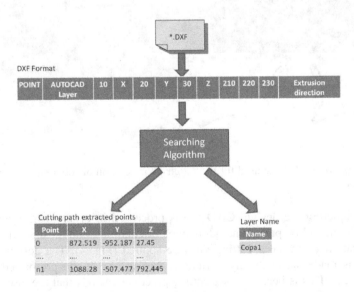

Fig. 5. Searching algorithm to extract the cutting point sequence

Finally, these points extracted from CAD model are used to obtain an B-spline approximation. B-spline curves and surfaces are a computationally efficient method of representing free-form shapes because of their capacity to approximate complex shapes through curve fitting and interactive curve design. [14].

Spline algorithm filters the orientation of the cutting path to smooth the cutting process, avoiding irregularities arising from sharp movements produce by the robot arm.

A B-spline formulation for a single segment can be written as:

$$S_i(t) = \sum_{k=0}^{3} P_{i-3+k}\, b_{i-3+k,3}(t) \; ; t\epsilon[0,1] \,. \tag{1}$$

Where S_i denotes the i-th B-spline segment, P denotes the set of control points, i and k denotes the local control point index, $b_{i,n}$ denotes a linear combination of order n and t_i are the control knots.

Considering cubic B-splines with uniform knot vector, linear combination of B-spline can easily be precalculated, and is equal for each segment in this case. Thus, uniform cubic B-spline segment can be written as:

$$S_i(t) = [t^3 \quad t^2 \quad t \quad 1]\frac{1}{6}\begin{bmatrix} -1 & 3 & -3 & 1 \\ 3 & -6 & 3 & 0 \\ -3 & 0 & 3 & 0 \\ 1 & 4 & 1 & 0 \end{bmatrix}\begin{bmatrix} P_{i-1} \\ P_i \\ P_{i+1} \\ P_{i+2} \end{bmatrix} \; ; t\epsilon[0,1] \,. \tag{2}$$

Thus, using the points extracted from CAD environment and applying the B-spline approximation, software application allows extracting curve fitting of the cutting path in order to generate the cutting robot routine. Fig. 6 and Fig. 7 show the result comparing B-spline approximation and extracted raw data from CAD environment.

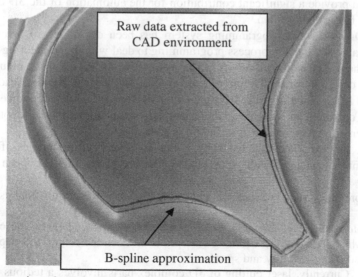

Fig. 6. Comparison between B-spline approximation and raw data extracted from CAD environment

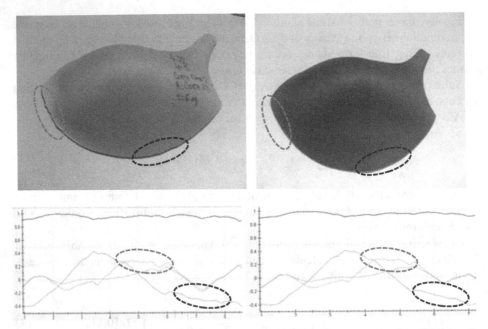

Fig. 7. Comparison of robot orientation (quaternion) with B-spline approximation (Right) and raw data extracted from CAD environment (Left)

3   Results and Discussion

The robotic cutting cell, based on laser technology, presented in this work was designed to provide a significant contribution for the automation of the 3D cutting of preformed fabric and/or foam.

A CAD-based robot programming system has been developed to automate 3D cutting. It allows a quickly process programming to deal with manufacturing changes, enabling the use of industrial robots in European Textiles and Clothing industry, that forces the design and manufacture of several Ready-to-wear collections each season (e.g. small batch manufacturing). Besides, the developed flexible robotic cell is easy to use. An operator can generate a robot routine, by simply drawing the robot paths on the CAD environment without specific training on robotics.

Finally, the experiments have shown that the laser cutting process fulfill the requirements of accuracy and repeatability demanded by the European Textiles industry.

Therefore, the system is expected to have the following impact for final users:

- *Increasing the European Textiles industry competitiveness:* The context in which Textiles and Clothing enterprises work in the future will depend even more on flexibility and speed.

    Currently, laser cutting of 3D complex parts involves a tedious manually program task (using teach pendant) due to the characteristics of laser processing to ensure orientation and stand-off distance with the surface of the part to be cut. This is why laser cutting of 3D bra cups is not cost-effective because programming task takes more than 2 hours for each model. The use of a CAD-based robot programming system reduces this tedious programming task to less than 1 minute because the operator only needs to load the CAD pattern of the part to be cut.

- *Increasing the technology level of the Textile sector:* Currently, 3D cutting process is a laborious semi-automatic cutting technique, based on the use of die cutters and subject to operator performance that challenge manufacturing efficiency and prevent the technological evolution of industrial processes. The developed CAD-based robot programming system allows the application of laser cutting technology for 3D preformed fabrics and/or foams applied to small batch manufacturing.
- *Increasing the productivity of the European Textiles sector:* Thus, to cut a bra cup, laser processing decreases cutting time from 30s (current semiautomatic mechanical cutting process) to 18s.

    Besides, the system can be easily reconfigured for small bath manufacturing. At this moment, manufacturing changes in the part to be cut involve an additional manual operation to change the model/shape of the die cutter to adjust to new designs. The use of an intuitive robot programming based on CAD eliminates this operation because manufacturing changes only involves loading of the new CAD pattern of the specific foam and/or preformed fabric to be cut.
- *Improving the quality and accuracy of the final part* enabling new cutting applications and product designs.

Therefore, this work yields a success example to encourage the development of robotics and laser technology focused on small batch manufacturing. Further efforts in this direction are needed to allow to SME to improve cost efficiency, performance, robustness and flexibility needed to reach a more competitive Textiles sector.

4   Future Work

Currently, robot processing routines are extracted directly from a CAD file. This involves the need to accurately represent the actual part to be cut. This requires a complex fixture system to avoid displacements of the part (e.g. preformed fabric).

Future work will address the development of a machine vision system to locate and align the part in robot coordinates. This will enable the adaptation of the robotic system to foam and fabric deformability, avoiding the use of demanding fixtures.

Acknowledgments. The work presented was supported by national funds from Fomento de la Investigación y la Innovación Empresarial program with code IN841C 2010/31 with the use of ERDF (European Regional Development Fund).

References

1. Eurostat, http://epp.eurostat.ec.europa.eu/cache/ITY_OFFPUB/KS-BW-09-001-04/EN/KS-BW-09-001-04-EN.PDF
2. Europe Innova, http://www.europe-innova.eu/c/document_library/get_file?folderId=24913&name=DLFE-2664.pdf
3. Payne, J.: Cutting Through the Surface: The Use of Laser Cutting Technology with traditional Textile Process. In: Textile Society of America Symposium Proceedings, Paper 43 (2010)
4. Walter, L., Kartsounis, G.A., Carosio, S.: Transforming Clothing Production Into a Demand-driven, Knowledge-based, High-tech Industry: The Leapfrog Paradigm. Springer (2009)
5. Neto, P., Pires, J.N., Moreira, A.P.: CAD-based off-line robot programming. In: 2010 IEEE Conference on Robotics Automation and Mechatronics (RAM), pp. 516–521 (2010)
6. Álvarez, M., Vidal, F., Iglesias, I., González, R., Alonso, C., Remuinan, M.: Development of a flexible and adaptive robotic cell for marine nozzles processing. In: 17th IEEE Conference on Emerging Technologies & Factory Automation (2012)
7. Neto, P., Mendes, N., Araújo, R., Pires, J.N., Moreira, A.P.: High-level robot programming based on CAD: dealing with unpredictable environments. Industrial Robot, Emerald 39(3), 294–303 (2012)
8. United Nations Industrial Development Organization, http://institute.unido.org/documents/M8_LearningResources/ICS/10.%20Laser%20technologies-%20a%20step%20forward%20for%20small%20and%20medium%20enterprises.pdf
9. Billard, A., Calinon, S., Dillmann, R., Schaal, S.: Robot Programming by Demonstration. In: Handbook of Robotics, ch. 59 (2007)
10. Argall, B.D., Chernova, S., Veloso, M., Browning, B.: A survey of robot learning from demonstration. Robotics and Autonomous Systems 57(5), 469–483 (2009)
11. Johnson, C.G., Marsh, D.: Modelling robot manipulators in a CAD environment using B-splines. In: IEEE International Joint Symposia on Intelligence and Systems, pp. 194–201. IEEE (1996)
12. Marri, H.B., Gunasekaran, A., Grieve, R.J.: An investigation into the implementation of computer integrated manufacturing in small and medium enterprises. The International Journal of Advanced Manufacturing Technology 14(12), 935–942 (1998)
13. AUTODESK, http://usa.autodesk.com/adsk/servlet/item?siteID=123112&id=12272454&linkID=10809853
14. de Boor, C.: A Practical Guide to Splines. Applied Mathematical Sciences, vol. 27. Springer (1978)

Haptic Guidance in a Collaborative Robotic System

Fernando Ribeiro[1] and António Mendes Lopes[2]

[1] Institute of Mechanical Engineering, LAETA
Faculty of Engineering, University of Porto
Rua Dr. Roberto Frias, s/n 4200-465 Porto, Portugal
[2] Institute of Mechanical Engineering, UISPA
Faculty of Engineering, University of Porto
Rua Dr. Roberto Frias, 4200-465, Porto, Portugal
aml@fe.up.pt

Abstract. The development of the notion of virtual fixture originated the broader concept of robot haptic guidance, meaning a recent technology to support motor learning. It has been applied in many areas, namely on automotive assembly, medical rehabilitation and training of healthy people. The implementation of virtual fixtures depends on the robot mechanical and drive system, namely the type of actuators and transmissions. If non-backdriven transmissions are used the operator controls the robot through the forces he / she exerts on the robot handle (which are measured by a force transducer incorporated at the robot end-effector). These robots usually use motors with large reduction ratios and are admittance controlled. In this paper we implement haptic guidance in an impedance type three degree-of-freedom (dof) heavy robot. An admittance low-level controller is firstly designed based on an IP (Integral, Proportional) velocity controller. Two types of virtual fixtures are implemented and the effectiveness of the proposed approach is illustrated experimentally.

Keywords: Collaborative robotic system, virtual fixtures, admittance control.

1 Introduction

In a collaborative robotic system a human operator and a robot interact in real-time while performing a given task [1]. This interaction pursues to replicate the physical movements (upper and / or lower limbs, hands or fingers) of the operator, possibly through some kind of mapping between the operator and robot workspaces. Usually, the term collaborative, applied to robotic systems, distinguishes between two models: cooperative manipulation and telemanipulation.

In telemanipulation there is an indirect interaction between the operator and the robotic system. The operator interacts with a local joystick (master system) to control a remote robot (slave system), responsible for the execution of the task. In cooperative manipulation the operator is physically positioned in the robot's workspace and interacts directly with it via some type of handler. Collaborative systems have been developed in addition to industrial robotic systems, which could not act in

P. Neto and A.P. Moreira (Eds.): WRSM 2013, CCIS 371, pp. 101–112, 2013.

coordination with humans for safety reasons. Moving the robot directly in its workspace has the advantage of keeping the operator kinaesthesia (the sensation of movement or strain in muscles, tendons, and joints), making cooperative manipulation more intuitive than telemanipulation. In both cases, there is some level of shared control between the human operator and the robot [1]. Sharing control through haptics implies that the operator experiences additional forces via the control interface he / she is grabbing to control the system [9]. Those forces can be repulsive, if they are used to create forbidden regions in the robot workspace [10], meaning that when the operator is closer to the boundaries of these regions, the higher the repulsive forces grow into. On the other hand, the feedback forces can be used to keep the operator on a given optimal (programmed) trajectory [11], being, in this case, attractive forces.

It is important to distinguish these two models of human-machine interaction (cooperative manipulation and telemanipulation) from the machine-machine interaction, usually called "cooperative systems", which denotes collaboration between two or more robots [19].

The concept of virtual fixture corresponds to overlapping abstract sensory information in a given workspace for the purpose of conditioning the system response [1-7]. This conditioning applied to collaborative systems generates a sensory stimulation on the operator in order to facilitate the execution of a task. This concept is rather broad and assumes that sensory information is not only positional, but also haptic (feeling of accelerations), visual and / or hearing.

An important application of virtual fixtures is assistive assembly, especially when dealing with heavy and large objects [1]. One solution would be to implement virtual points, curves, surfaces or volumes, such as a funnel in 3D space to guide the parts to a given desired point in the workspace. This virtual funnel would have two purposes: to avoid collisions during the movement and guide the operator and the load to the mounting location. The virtual fixtures are not limited to simulate rigid (solid) environments, but can also simulate surfaces with lower stiffness (ex. linear / non-linear spring type), or repulsive / attractive surfaces and friction contacts [1-8].

The implementation of virtual fixtures depends on the robot mechanical and drive system, namely the type of actuators and transmissions. If backdriven transmissions are used, the robot actuators apply forces to the operator, whenever he / she tends to violate a virtual restriction, for example. Such robots typically use direct drive motors or small reductions. Non-backdriven transmissions allow the operator to control the robot through the forces he / she exerts on the robot handle (which are measured by a force transducer incorporated at the robot end-effector). These robots usually use motors with large reduction ratios. In the former case, the robot will be impedance controlled and in the latter case it will use admittance control [1, 3].

Several pioneer works [1-8] reporting the use of virtual fixtures are, for example, Abbot et al. [3] that analyze virtual fixtures in the context of both cooperative manipulation and telemanipulation systems, considering issues related to stability, passivity, human modeling and applications. They present the design, analysis, and implementation of two categories of virtual fixtures: guidance virtual fixtures, which assist the user in moving the manipulator along desired paths, and forbidden-region virtual fixtures, which prevent the manipulator from entering into forbidden regions of the robot workspace. Taylor et al. [8] propose a robotic system designed to extend the operator ability to perform small-scale manipulation tasks. In their approach, the tools are held, simultaneously, by the operator and the robot. Forces exerted by the operator on the tool and by the tool on the environment are measured and used by the controller to offer smooth, tremor-free and precise positional control and force scaling. Payandeh and Stanisic [4] use virtual fixtures in telemanipulation and training environments. They found that virtual fixtures could improve the speed and precision of the operator, reduce the operator workload and the duration of the training phase for novice operators. Many studies can also be referred dealing with virtual fixtures on cooperative manipulation and telemanipulation systems of both impedance and admittance types [4-7].

During the last two decades, the development of the notion of virtual fixture originated the broader concept of robot haptic guidance, meaning a recent technology to support motor learning. It has been applied in many areas, namely medical rehabilitation [12-13] and training of healthy people [14, 18]. For example, Marchal-Crespo and Reinkensmeyer [12] review control strategies for robotic therapy devices. Several strategies have been proposed, including assistive, challenge-based, haptic simulation and coaching. Takahashi et al. [15] investigate how a robot can improve motor function, concluding that robot based therapy yields improvements in hand motor function after chronic stroke. Emken and Reinkensmeyer [16] studied robot enhanced motor learning in human locomotion. They concluded that motor learning of a novel dynamic environment can be accelerated by exploiting the error based learning mechanism of internal model formation. Reinkensmeyer and Patton [17] show how robotic devices can temporarily alter task dynamics in ways that contribute to motor learning experience, suggesting possible applications in rehabilitation and sports training. Ben-Pazi et al. [14] investigate the effect of mechanical properties of a pen on the quality of handwriting in children. A pen was attached to a robot and effective weight (inertia) and viscosity were programmed. Increased inertia and viscosity improved handwriting quality in 85% of children. Abbink et al. [9] argue that haptic shared control is a promising approach to meet the commonly voiced design guidelines for human-automation interaction, especially for automotive applications.

In this paper we implement haptic guidance in an impedance type three degree-of-freedom (dof) heavy robot. An admittance low-level controller is firstly designed based on an IP (Integral, Proportional) velocity controller. Two types of virtual fixtures are implemented and the effectiveness of the proposed approach is illustrated experimentally. Bearing these ideas in mind, the paper is organized as follows. In section 2 the used robot is introduced and the admittance controller is presented. Section 3 describes the virtual fixtures, its implementation and testing. Finally, in section 4, the main conclusions are presented.

2   Robotic System and Low-Level Admittance Controller

The robotic system consists of an existing Cartesian manipulator and a PC-based digital controller. This robot was not designed for human interaction; it was primarily designed to carry heavy loads at high speeds. It has three linear axes powered by brushless AC servomotors. Ball-screw based transmissions convert the motors rotation into linear motion. The axes linear position and acceleration are monitored via incremental encoders and accelerometers, respectively. A 6-axis force/torque transducer is mounted between the robot end-point and the operator handle. The torque signals were not used. The controller runs under Matlab / Simulink / xPC Target.

Each robot axis may be modeled as a mass-damper system and its dynamics is approximately given by:

$$T = K_T \cdot I \tag{1}$$

$$T = J\ddot{\theta} + B\dot{\theta} \tag{2}$$

$$\theta = K_{LA} \cdot x \tag{3}$$

The drive current, I, for the motor produces the torque, T, at the motor shaft, which is rigidly connected to the load. Constant J is the total inertia of the parts and B the total friction (essentially viscous) both referred to the motor shaft. Parameter K_T is the torque constant and K_{LA} represents the transmission ratio.

We implemented an IP velocity controller that includes an integrator anti-windup loop. This loop prevents the integrator to saturate by adding feedback of the error (the difference between the actuator output and the control action signal) multiplied by a tracking time constant. If the integrator saturates, adding the error signal will reset it.

The admittance controller relies on the velocity controller and a linear relationship between the force imposed by the operator and the robot velocity, as given by (Figure 1):

$$\dot{x} = c \cdot f \tag{4}$$

Where $c > 0$ is an admittance gain that acts like the inverse of a damping coefficient. Thus, the admittance controller transfer function, $G_A(s)$, is given by:

$$G_A(s) = \frac{sX(s)}{F(s)} = \frac{c}{s^2 \dfrac{J \cdot K_{LA}}{K_I \cdot K_T} + s \dfrac{\left(B \cdot K_{LA} + K_T \cdot K_P\right)}{K_I \cdot K_T} + 1} = \frac{c}{\dfrac{s^2}{\omega_n^2} + \dfrac{2\xi}{\omega_n} s + 1} \tag{5}$$

It should be noted that c acts as steady state gain that does not affect the setting of the parameters of the velocity controller. Moreover, the natural frequency of the controlled system was chosen equal to $\omega_n = 60$ rad/sec, based on simulations and experiments involving velocity / admittance control, and the damping coefficient was set to $\xi = 1$, fixing the controller gains, K_I and K_P, respectively. The values of all parameters are shown in Table 1.

In Figure 2 the response to an arbitrary force command imposed by the operator is shown ($\omega_n = 60$ rad/s; $\xi = 1$; $c = 1$ m·s^{-1}/N). It can be seen that the robot responds to the force profile with a velocity. As expected, force and velocity are numerically equal, as the impedance gain, in this case, was set equal to unity.

Table 1. Values of the system and controller parameters

Parameter	Value
K_T	0.590 N · m/A
	(from manufacturer catalog)
J	axis X: 2.57×10^{-3} Kg · m²; axis Y: 1.77×10^{-3} Kg · m²; axis Z: 1.04×10^{-3} Kg · m²
	(calculated, taking into account axes inertias and masses)
B	0.002 N · m · s/rad
	(0.001 N · m · s/rad, from manufacturer catalog, plus 0.001 N · m · s/rad, estimated to correspond to approximately 10% of the available torque at maximum speed)
K_{LA}	axis X: 314 rad/m; axis Y: 314 rad/m; axis Z: 157 rad/m
	(from manufacturer catalog)
ω_n	60 rad/s
	(adjusted by simulation)
ξ	1
	(adjusted in order to have no overshoot)

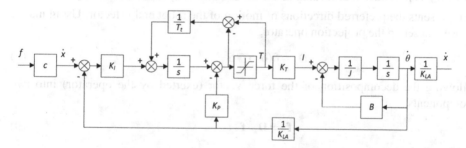

Fig. 1. Block diagram of the admittance controller

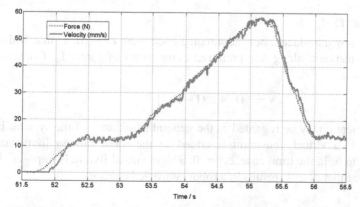

Fig. 2. Response to a given force command imposed by the operator
($\omega_n = 60$ rad/s; $\xi = 1$; $c = 1$ m · s⁻¹/N)

3   High-Level Virtual Fixtures

Generally speaking, adopting an admittance controller, we establish the relationship between force imposed by the operator and motion of the robot end-effector, as given by [5]:

$$v = \Phi(f) \tag{6}$$

Where $f \in \Re^3$ is the force imposed by the user and $v \in \Re^3$ is the velocity of the robot end-effector, both expressed in the Cartesian space. The admittance function Φ establishes the relationship between f and v. If this relationship is linear and the same in all directions, we can write:

$$v = c \cdot f \tag{7}$$

Thus, it is understood that the velocity of the robot in a given direction is proportional to the force exerted in that direction and the robot has an isotropic behavior in terms of velocity.

A virtual fixture generalizes the previous model by adding anisotropy conditions to the robot workspace. For doing this, the time dependent $3 \times n$ $(0 < n < 3)$ matrix $\delta = \delta(t)$ is introduced, according to the notation used by Bettini et al. [5]. Intuitively, δ represents the preferred directions of motion of the robot end-effector. Using matrix δ, we can set up the projection operator,

$$D_\delta = \delta(\delta^T \delta)^{-1} \delta^T \tag{8}$$

allowing the decomposition of the force vector (exerted by the operator) into two components,

$$f_\delta = D_\delta \cdot f \tag{9}$$

$$f_\tau = f - f_\delta \tag{10}$$

meaning that $f_\tau^T \cdot f_\delta = 0$.

We can now introduce a new admittance coefficient $c_\tau \in [0, 1]$ that will attenuate the system response along non-preferred components of force, f_τ. Consequently, it results in:

$$v = c[D_\delta + c_\tau(I - D_\delta)] \cdot f \tag{11}$$

The coefficient c may be regarded as the general admittance of the system. Imposing $0 < c_\tau \leq 1$, a virtual constraint is added to the robot motion in the directions orthogonal to δ. In the limit case, $c_\tau = 0$, a rigid virtual fixture is imposed. It should also be noted that $c_\tau = 1$, results in a robot isotropic behavior.

4   Experimental Results

In this section two different situations are implemented, using the Cartesian robot and low-level admittance controller described in section 2. Afterwards, the experimental results are discussed.

4.1   Motion Along a Curve

In this case the robot end-effector can be moved along a curve. It is assumed that the virtual constraint is given by the parametric expression [5]:

$$\mathbf{p}(s) = [x(s) \quad y(s) \quad z(s)]^T, s \in [0,1] \tag{12}$$

Defining $\mathbf{p}[s(\mathbf{x}_a)]$ as the point of the virtual curve closer to the robot end-effector real position, $\mathbf{x}_a$, the preferred direction of motion δ is given by the normalized vector tangent to the curve at that point:

$$\mathbf{t}(\mathbf{x}_a) = \frac{d}{ds}\mathbf{p}(s)\Big|_{s=s(\mathbf{x}_a)} \tag{13}$$

$$\delta(\mathbf{x}_a) = \frac{\mathbf{t}(\mathbf{x}_a)}{\|\mathbf{t}(\mathbf{x}_a)\|} \tag{14}$$

Nevertheless, if the robot end-effector does not start on the desired curve it will tend to move along a direction parallel to the curve. This means that an attractor must be defined to redirect the robot to the desired path. This can be done using:

$$\delta_c(\mathbf{x}_a) = \text{signal}[\mathbf{f} \cdot \delta(\mathbf{x}_a)] \cdot \delta(\mathbf{x}_a) + k_d \mathbf{e}(\mathbf{x}_a) \tag{15}$$

$$\mathbf{e}(\mathbf{x}_a) = \mathbf{p}[s(\mathbf{x}_a)] - \mathbf{x}_a \tag{16}$$

Where the $\mathbf{e}(\mathbf{x}_a)$ is the Cartesian error and k_d the parameter that controls the rate of convergence to the desired curve.

Figure 3a illustrates the robot end-effector being moved by the user and guided along a curve. In this case, a helix curve is defined as the desired path, as given by:

$$x(s) = x_c + r\cos(2\pi s) \tag{17}$$

$$y(s) = y_c + r\sin(2\pi s) \tag{18}$$

$$z(s) = z_c - 2\pi b s \tag{19}$$

With center $\mathbf{x}_c = [x_c \; y_c \; z_c]^T = [0 \; 0 \; 0]^T$, radius $r = 100$ mm and pitch $b = 15$ mm. The direction $\delta(s)$ in every point, s, is given by:

$$\delta(s) = \frac{[\dot{x}(s) \quad \dot{y}(s) \quad \dot{z}(s)]^T}{\|[\dot{x}(s) \quad \dot{y}(s) \quad \dot{z}(s)]^T\|} \tag{20}$$

The control parameters $c = 1$ m $\cdot$ s^{-1}/N and $c_\tau = 0$ m $\cdot$ s^{-1}/N are used.

As can be seen, while at the beginning the robot end-effector is away from the desired path ($\mathbf{x}_i = [0\ 0\ 0]^T$), it rapidly converges to the curve and stays approximately on there. Figure 3b depicts the modulus of the error. We can see the approach phase, on the left of the graph, where the error diminishes quickly. Subsequently, the error is kept inferior to 0.6 mm.

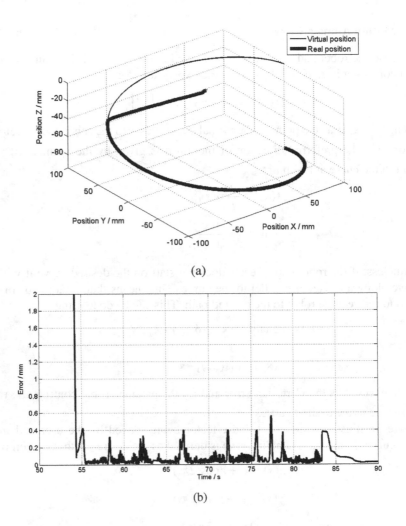

(a)

(b)

Fig. 3. (a) Motion along a helix curve; (b) time evolution of the error

4.2   Motion Inside a Tube

With this type of restriction the robot end-effector can be freely moved inside the tube. Once in the tube, it must stay in there. It this case the task can be described by a parametric curve $\mathbf{p}(s)$ representing the axis of a tube with radius r_t. The boundary surface of the tube is a switching surface between a free motion region (inside) and a virtual attractive region. Further, we define a transition region $\varepsilon_0 < \varepsilon < r_t$ within which the gain discontinuity is smoothed [5].

$$c_{tu} = \begin{cases} c_\tau, & \|\mathbf{e}(\mathbf{x}_a)\| > r_t \\ c_\tau - \left[\dfrac{r_t - \|\mathbf{e}(\mathbf{x}_a)\|}{\varepsilon}\right]^n \cdot (c_\tau - 1), & (r_t - \varepsilon < \|\mathbf{e}(\mathbf{x}_a)\| \le r_t \wedge [\mathbf{e}(\mathbf{x}_a)\cdot\mathbf{f} < 0] \\ 1, & \text{all other cases} \end{cases} \tag{21}$$

Where $n \ge 1$ is a scalar that shapes the switching surface. If the end-effector is outside the tube the surface is virtually attractive. If it is inside the tube the surface is virtually repulsive.

The reference direction is given by

$$\delta_{tu}(\mathbf{x}_a) = \text{signal}[\mathbf{f} \cdot \delta(\mathbf{x}_a)] \cdot \delta(\mathbf{x}_a) + k_d \mathbf{e}_t(\mathbf{x}_a) \tag{22}$$

$$\mathbf{e}_t(\mathbf{x}_a) = \begin{cases} 0, & \|\mathbf{e}(\mathbf{x}_a)\| < r_t \\ \dfrac{\mathbf{e}(\mathbf{x}_a)}{\|\mathbf{e}(\mathbf{x}_a)\|}[\mathbf{e}(\mathbf{x}_a) - r_t], & \|\mathbf{e}(\mathbf{x}_a)\| \ge r_t \end{cases} \tag{23}$$

We used the helix defined in the previous example for the axis of the tube. The tube radius was set to $r_t = 12$ mm and the control parameters are $c = 1$ m·s^{-1}/N, $c_\tau = 0$ m·s^{-1}/N, $n = 3$ and $\varepsilon = 10$ mm. Figure 4a shows the tube and the robot trajectory. It can be seen that the robot starts outside the tube ($\mathbf{x}_i = [0\ 0\ 0]^T$) and rapidly converges to the inside volume. Once there, the robot can be freely moved, explaining the almost "random" trajectory observed in graph. Figure 4b depicts the modulus of the distance between the robot end-effector and the axis of the tube. As in the previous example, we can see the approach phase, on the left of the graph, where the distance diminishes quickly. Subsequently, the distance is at most 12 mm, approximately, meaning that the end-effector is close to the tube surface, but always inside it.

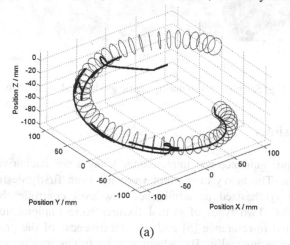

(a)

Fig. 4. (a) Motion inside a tube; (b) time evolution of the error

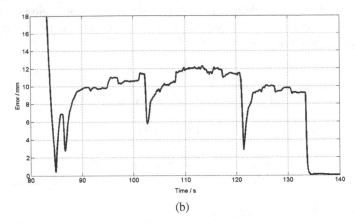

(b)

Fig. 4. (*Continued*)

Figure 5a depicts the robot handle actuated by the operator. In Figures 5b and 5c an example of haptic guided motion is illustrated. The desired trajectory is a circumference, being executed without assistance (Figure 5b) and with assistance (Figure 5c). As can be seen, while in the former case it is almost impossible to keep the robot on the desired trajectory, in the latter case the trajectory is easily executed.

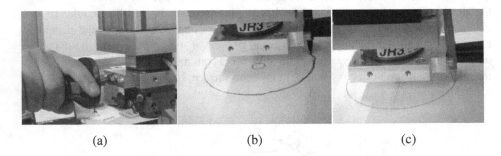

(a)   (b)   (c)

Fig. 5. Images taken during haptic guidance along a circle

5   Conclusions

In this paper haptic guidance was implemented using a non-backdriven three degree-of-freedom robot. The heavy duty robot were not been firstly designed for human interaction. We synthesized an admittance low-level controller based on an IP velocity controller. Two types of virtual fixtures were implemented based on the formalism adopted in reference [5] and the effectiveness of the proposed approach was illustrated experimentally. The admittance controller was proven to satisfy the requirements imposed by the implementation of virtual guidance control: the human can control the robot in a virtualized environment that restricts or aids the human arm motion.

References

1. Peshkin, M.A., Colgate, J.E., Wannasuphoprasit, W., Moore, C.A., Gillespie, R.B., Akella, P.: Cobot Architecture. IEEE Transactions on Robotic and Automation 17(4), 337–390 (2001)
2. Rosenberg, L.B.: Virtual Fixtures: Perceptual Tools for Telerobotic Manipulation. In: Proc. IEEE Virtual Reality Int. Symp (VRAIS 1993), pp. 76–82 (1993)
3. Abbott, J.J., Marayong, P., Okamura, A.M.: Haptic Virtual Fixtures for Robot-Assisted Manipulation. Springer Tracts in Advanced Robotics 28, 49–64 (2007)
4. Payandeh, S., Stanisic, Z.: On application of virtual fixtures as an aid for telemanipulation and training. In: Proc. 10th Symposium on Haptic Interfaces for Virtual Environments and Teleoperator Systems, pp. 18–23 (2002)
5. Bettini, A., Marayong, P., Lang, S., Okamura, A.M., Hager, G.D.: Vision-assisted control for manipulation using virtual fixtures. IEEE Trans. Robotics 20(6), 953–966 (2004)
6. Park, S., Howe, R.D., Torchiana, D.F.: Virtual fixtures for robotic cardiac surgery. In: Proc. 4th Int. Conf. on Medical Image Computing and Computer-Assisted Intervention, pp. 1419–1420 (2001)
7. Turro, N., Khatib, O.: Haptically augmented teleoperation. In: Rus, D., Singh, S. (eds.) Experimental Robotics VII. LNCIS, vol. 271, pp. 1–10. Springer, Heidelberg (2001)
8. Taylor, R., Jensen, P., Whitcomb, L., Barnes, A., Kumar, R., Stoianovici, D., Gupta, P., Wang, Z., Dejuan, E., Kavoussi, L.: Steady-hand robotic system for microsurgical augmentation. Int. J. Robotics Research 18(12), 1201–1210 (1999)
9. Abbink, D.A., Mulder, M., Boer, E.R.: Haptic shared control: smoothly shifting control authority? Cogn. Tech. Work 14(1), 19–28 (2012)
10. Marayong, P., Okamura, A.M.: Speed-accuracy characteristics of human-machine cooperative manipulation using virtual fixtures with variable admittance. Hum. Factors 46(3), 518–532 (2004)
11. Griffiths, P.G., Gillespie, R.B.: Sharing control between humans and automation using haptic interface: primary and secondary task performance benefits. Hum. Factors 47(3), 574–590 (2005)
12. Marchal-Crespo, L., Reinkensmeyer, D.J.: Review of control strategies for robotic movement training after neurologic injury. Journal of Neuroengineering and Rehabilitation 6(20) (2009)
13. Kahn, L.E., Zygman, M.L., Rymer, W.Z., Reinkensmeyer, D.J.: Robot-assisted reaching exercise promotes arm movement recovery in chronic hemiparetic stroke: A randomized controlled pilot study. Journal of Neuroengineering and Rehabilitation 3(12) (2006)
14. Ben-Pazi, H., Ishihara, A., Kukke, S., Ranger, T.D.: Increasing viscosity and inertia using a robotically controlled pen improves handwriting in children. Journal of Child Neurology 25(6), 674–680 (2009)
15. Takahashi, C.D., Der-Yeghiaian, L., Le, V., Motiwala, R.R., Cramer, S.C.: Robot-based hand motor therapy after stroke. Brain 131(2), 425–437 (2007)

16. Emken, J.L., Reinkensmeyer, D.J.: Robot-enhanced motor learning: Accelerating internal model formation during locomotion by transient dynamic amplification. IEEE Trans. on Neural Systems and Rehabilitation Engineering 13(1), 33–39 (2005)

17. Reinkensmeyer, D.J., Patton, J.L.: Can robots help the learning of skilled actions? Exercise and Sport Sciences Reviews 37(1), 43–51 (2009)

18. Lüttgen, J., Heuer, H.: The influence of haptic guidance on the production of spatio-temporal patterns. Human Movement Science 31(3), 519–528 (2012)

19. Marayong, P.: Motion Control Methods for Human-Machine Cooperative Systems. PhD Thesis, John Hopkins University (2007)

Tailor Made Robot Co Workers
Based on a Plug&Produce Framework

Andreas Pichler[1], Paolo Barattini[2], Claire Morand[3], Ibrahim Almajai[4],
Neil Robertson[3], James Hopgood[4], Paolo Ferrara[5], Matteo Bonasso[2],
Christoph Strassmair[6], Maren Rottenbacher[6], Max Staehr[6], Rüdiger Neumann[7],
Manuel Tornari[8], Alberto Rovetta[8], Matthias Plasch[1], Herald Bauer[1],
and Christian Wögerer[1]

[1] Robotics and Adaptive Systems
PROFACTOR GmbH
Steyr-Gleink, A-4407, Austria
Christian.woegerer@profactor.at
[2] Ridgeback s.a.s.
via S. Francesco da Paola 6
Torino, 10123, Italy
[3] School of Engineering and Physical Sciences
Heriot-Watt University
Edinburgh, EH14 4AS, Scotland, United Kingdom
[4] School of Engineering
The University of Edinburgh
Edinburgh, EH9 3JL, Scotland, United Kingdom
[5] FerRobotics Compliant Robot Technology GmbH
Stock, Science Park
Linz, A-4040, Austria
[6] Institute for Applied Research
Ingolstadt University of Applied Sciences
Ingolstadt, 85049, Germany
[7] Festo AG + Co.
Esslingen am Neckar, 3373734, Germany
[8] Laboratory of Robotics
Politecnico di Milano
Milano, 20156, Italy

Abstract. Current industrial robots are optimized to "economy of scale" while new business models demand customization, individualization and service-orientation. The next generation of robotic workers will have to cope with more complex tasks, rapidly adapt to new situations and provide high flexibility. Additionally demographic change requires a paradigm shift to put the human workforce in the Center of a production system. The LOCOBOT project proposes a toolkit for building customized low cost robot co-workers for a broad spectrum of scenarios. The system approach envisaged keeps the human worker in the loop. So far the approach is user centred design.

Keywords: Plug&Produce, Mobile Platform, Framework.

P. Neto and A.P. Moreira (Eds.): WRSM 2013, CCIS 371, pp. 113–126, 2013.
© Springer-Verlag Berlin Heidelberg 2013

1   Introduction

1.1   Problem Description

The ultimate goal of LOCOBOT is to provide safe, tailor made low cost systems which are socially accepted by the worker and enable greener, more customized and high quality products, capable of relieving the human worker from part of the tasks till now not automatized, as well easy adaptability. The paper presents a framework which allows designing complex systems based on ergonomics and human robot interaction requirements out of a modular robot tool kit. Specifically we will describe how individual patterns of interaction affect and evolve the design of a robot co-worker in a manufacturing environment.

1.2   State of the Art

In the manufacturing plants the existing robots use rigid and heavy structures based on electrical motors, causing safety problems in human- robot cooperation, and their control programs need to be manually adjusted and programmed for given kinematics, and tasks have to be defined in great detail. Human Robot Interaction for close cooperation is still neglected.

2   Detailed Description of Scenario

The LOCOBOT consists of an autonomous platform with a resilient arm, a self-adapting Finray soft gripper, and multiple sensors for safety and HRI, a set of software modules.

The project developed HRI and Technical requirements on the basis of three scenarios in a car assembly line. They include physical cooperation and task sharing with the Human worker that has supervisory role. In these demonstrators the LOCOBOT relieves the human from most of the physical work.

2.1   Demonstrator 1 – Starter Presorting

We would like to encourage you to list your keywords in this section. In this scenario [1] the LOCOBOT has to locate a kart containing the starter of a certain model, scan the content, identify the starter, pick it up and bring it to the orange kart. It needs HRI in case of malposition of the part, presence of paper board sheets between different layers of starters, absence of part/kart/rack, failure in identification of the part, unidentifiable obstacle occupying the work space. Currently this work is done completely manually by the Human worker.

2.2   Demonstrator 2 – Battery Pre-picking

In this case car batteries are delivered to the work area on pallets. The LOCOBOT has to identify the pile of the requested battery model, pick up a battery and bring it to a

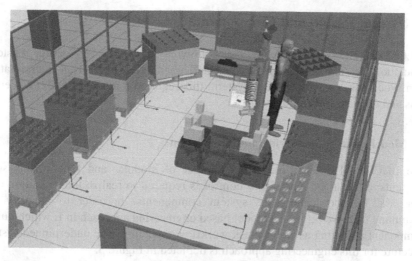

Fig. 1. Demonstrator 2

wheeled kart. The HRI is needed for the lifting of the battery handle (so that can be gripped), removing cardboard layers, malposition of the battery etc. Currently this work is done completely manually by the Human worker.

2.3   Demonstrator 3 – Battery Mounting

The car batteries are picked in the right sequence from a three lane slide-rack. Then the LOCOBOT moves to the car that is in the assembly, moved by the lifting crane at slow speed. The LOCOBOT has to track the car speed and present the battery at the correct position inside the hood. The battery will be positioned in collaboration with the human worker.

3   Ergonomics and Human Robot Interaction Addressing the Needs of Individual Human Workforce

The LOCOBOT relieves the human of most of the physical load and repetitive movements in the 3 demonstrators. The ergonomics evaluation was performed using NIOSH and OCRA methods in order to assess the improvement in workload for human worker, but the new situations arising from the interaction with LOCOBOT. The assessment showed positive results.

Nevertheless the apparently simple tasks the HRI in the industrial Environment requested a great design effort to match the Regulatory constraints, the cluttered and noisy environment, and the lack of standards for Autonomous Robots feedback, alarm signals, coded vocal and gesture commands [2]. The HRI design included vocal and gesture commands, visual and sound feedbacks. Sometimes the HRI design had to come to terms to other engineering requirements, for example to avoid the interference of blinking LED light for visual feedback to the user versus the different sensors systems and cameras.

4   System Design and Components

The LOCOBOT system includes the following modules: Navigation System, Manipulation System (LocoArm), Object-Recognition, Interaction-Interpretation System (vocal and gesture), Safety System, this required a model based engineering process and relevant Integration effort.

4.1   Model Based Engineering and Simulation

As the LOCOBOT is composed of a number of modular and functional system components, a means of supervisory control is required to realize the execution of a certain task, by coordinating the system components. In order to simplify the programming task, we propose a model based engineering approach in E which builds on component modeling as well as task workflow modeling. The underpinned system architecture for this engineering approach is depicted in Figure 3.

In the first step, the LOCOBOT composed out of its system components is modeled in the 3D simulation environment 3DCreate (link to VIS). Moreover, the geometry models are enhanced with a behavior description. This description abstracts the services, the component provides to the LOCOBOT system. The behavior description is expressed through service functions and corresponding input and output parameters. After the LOCOBOT system has been modeled (see Figure 2), the robot assistance system is programmed by simply expressing the desired process, using a graphical workflow. A workflow editor allows access to the component behavior descriptions of the 3D simulation environment [1] Based on the behavior information, a palette of workflow activities is created which can be used to model the process. Additionally a selection of control flow activities can be used to influence the execution order of the workflow activities, e.g. based on logical conditions.

The engineering step from the process workflow model to an executable control application is performed by a code generator which transforms the workflow model into an IEC 61499 compliant Function Block (FB) control application. Workflow activities are transformed into FB types based on the transformation rules of the generator's data model [ ] M. Based on the generated FB types, a FB network is generated by transforming the activity connections into representing event and data connections according to IEC 61499. The code generator builds on the 4DIAC framework, a tool-chain for modeling distributed control applications. Additionally it provides an IEC 61499 compliant runtime environment (FORTE), which is used to execute the generated control application.

The circle of the model based engineering process is closed by either running the control application on the real target system, i.e. accessing the real hardware/software system components, or by *testing* the application behavior with the aid of the 3D simulation environment. In order to simulate the behavior, the FB runtime environment is coupled with 3DCreate and the executing control application sends signals to the simulation environment, which can trigger movements of the components models. Comparable to real system components, the simulated components can emit signals to the control application in return, e.g. to report events or exceptions.

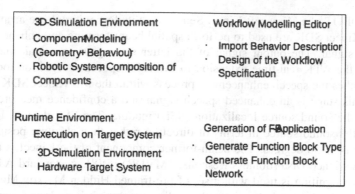

Fig. 2. System architecture of the engineering process

4.2   Human Robot Interaction

1) Distant Speech Recognition

Many autonomous robots do not implement human vocal commands, thereby lacking an enhanced HRI experience. LOCOBOT requires distant speech recognition (DSR) because it enables man-machine interaction through speech without the necessity of putting on intrusive body-mounted or head-mounted microphones. However, since LOCOBOT is also required to work in a factory environment, then the acoustic conditions are such that the acoustic signal-to-noise ratio (SNR) is very low, there are high reverberation levels, and there will be competing speech from co-workers. As a result, DSR is particularly challenging in this adverse acoustic environment.

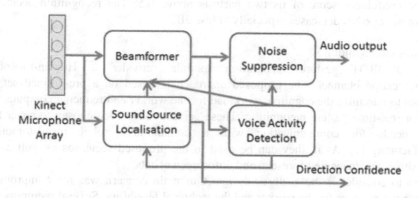

Fig. 3. Simplified Kinect Audio Processing

To achieve a reasonably efficient DSR, a microphone array with known geometry is utilized, which facilitates a variety of speech enhancement and audio localization methods. These techniques exploit the different time of arrival of speech at each channel in a microphone array, which can then be used to identify the target speech position, and to steer a beam-former in the direction of the talker thereby suppressing noise and competing speech and enhancing the target signal [5][6].

In LOCOBOT, the Microsoft Kinect Sensor with its four microphone array and the Microsoft Kinect SDK are used to perform spatial beam-forming and enhancement of the speech signal. The output signal of the latter stage goes to a real-time speech recognizer, the ATK real-time extension to the HTK speech recognition toolkit [7]. Fig. 5 describes the speech enhancement process within the MS Kinect SDK [8]. The output of this stage is an enhanced speech signal and a confidence measure in how accurate is the sound source localization (SSL) module. The SSL detects, localizes, and tracks the sound source in terms of direction rather than absolute position. The angle range is +/-50 degrees. The beam-former component of the Kinect is based on Minimum Variance Distortionless Response (MVDR) [5][8]. A standard ATK setup for speech recognition is used with a set of pre-trained Hidden Markov Models that comes with it. In this work, however, the audio capture unit is no longer needed and changes were made in the code to have the audio buffered by the output of the Kinect SDK beam-former. Cepstral Mean Subtraction is applied to the Mel Frequency Cepstral Coefficients features to better cope with channel characteristics. Confidence scores are based on recognition scores relative to a "background model" [7].

A preliminary performance evaluation is carried out by playing speech commands from one loudspeaker and white noise from the other. Both loudspeakers are 1m away from the Kinect sensor. The database used in the evaluation is composed of 19 commands uttered twice by a male speaker and recorded by close-talking microphone in an audio lab. The 38 command recordings were played through one loudspeaker at different SNRs. Each speech command is composed of two parts; the first part is the word robot and the second part is one these 19 commands such as START, STOP, SLOWER, etc. A recognized speech command is only taken into consideration if the average confidence score of its two parts is above 0.5. The recognition accuracy deteriorates as SNR decreases especially at low dB.

2) Gesture Recognition

The LOCOBOT gesture recognition module provides a Human-to-Robot communication channel. The proposed approach is based on a pre-defined set of gestures to minimize the cognitive workload for the worker and reducing computation load for real-time gesture recognition. These gestures have been chosen to cover the basic needs for communication with a collaborative robot in automotive manufacturing [2]. As it, they can be used in the proposed scenarios as well as in other situations, managing foreseen and unforeseen events.

First to consider is the gestures design. Our main concern was the compromise between the easiness for the worker and the technical feasibility. Several requirements have been identified. From the worker point of view, the main principles are (i) easy to make and remember (ii) use of single limb, natural, gesture (arm and/or hand and/or fingers) to let the second hand of the worker free for example to handle a tool (iii) the use of the right or left hand is not to be imposed. On a practical stand, the following rules have been considered: (i) Any worker in the area can command the robot but only one at a time for safety reasons. The worker identify himself with a specific gesture (called "identification") and release the robot with a "change" command, (ii) large, dynamic gestures are preferred for understandability at any distance, including far range (e.g. the "stop" command) (iii) combined hand movement and fingers pattern gestures are used for closer distance to the robot (e.g. a "report" status command). Examples of proposed gestures are presented in Figure 5 (a) and (b).

Second, we proposed a simple gesture recognition algorithm for real-time processing and dataset feasibility evaluation. It is implemented using a ©Kinect sensor as this device meets the LOCOBOT requirements of low cost component with decent quality. We choose a one-shot-learning approach, actively studied in the literature [7]. The advantage is that it allows easily defining and learning a new gesture when a new use-case scenario requires a specific command. A single example of each gesture in the data set is recorded and used as the gesture class model. During real-time interaction, a candidate gesture is segmented in the video stream by detecting passages to the resting position (hands lying along the body). The distance between this candidate gesture and each of the gesture representatives is computed. The gesture is recognized if (i) its distance to one of the model is less than a given threshold and (ii) it is the smallest distance to any other model. The feature used is the hands positions w.r.t. the head position. Indeed, the sequence of hand positions is the elementary common character between a wide variety of gestures: single and double handed dynamical gestures and single and double handed pose gestures, which can be furthermore combined with hand pattern gestures. These positions are obtained through a skeletal extraction method (provided by the Microsoft ©Kinect SDK, Figure 5 (c) and (d)). The distance measure chosen is based on the Dynamic Time Warping algorithm to cope with the fact that a gesture can be done at different paces by different workers.

An evaluation has been run on the whole 19 gestures constituting the proposed dataset. Six different users performed each gesture three times. Evaluation of the method shows low confusion and high accuracy. Gestures in the dataset can be grouped into three categories. Group recognition rates are: Dynamical 72%, postural 85.5% and two handed 100%. The intra group recognition rates are lower due to the fact that some gestures in the group have the same dynamic pattern and are distinguishable by fingers pattern. The mean intra-group recognition rate are Dynamical 67%, postural 37% and two handed 52%. Further improvement is to combine a finger pattern recognition algorithm to this hand position approach.

Fig. 4. Gesture recognition module. Example of gestures (a) dynamic, (b) postural (bottom) gesture recognition principle (c) skeleton extraction (d) normalized hand trajectory (from [1])

4.3   LOCOBOT Platform and Navigation

The LOCOBOT platform is a one volume frame vehicle with two sets of independent wheel complexes. Each wheel includes: an electrical motor, a gear head, a mechanic wheel and a hub which connects and supports weights and dynamics of the components. The mechanic wheel is a vectorial motion wheel: it has on its peripherals a series of free spinning rolls placed at a 45° to the wheels rotational axis, so that the vehicles direction and sense of movement can be controlled without a steering system.

To ensure that workers in assembly line can work and move around and interact with the LOCOBOT without risk, obstacle (including mobile objects) detection and avoidance system is implemented based on eight infrared and eight sonar sensors, two on each of its sides. If the object represents a danger, the robot slows down and stops in case of proximity. It automatically resumes the path when the obstacle has been removed.

The Human worker interacts using a GUI through the touch screen located on the robot or a remote control by PC, ipad and iphone or through vocal and gesture commands

The control system is based on the Robotic Operating System (ROS). The navigation system includes a Coarse Navigation layer and a Fine Navigation layer. Coarse Navigation has a SLAM approach enabling the robot to perform mapping using a LIDAR, positioning within to achieve autonomously the desired position, by calculating a trajectory to avoid obstacles along the path. The goal of Fine Navigation

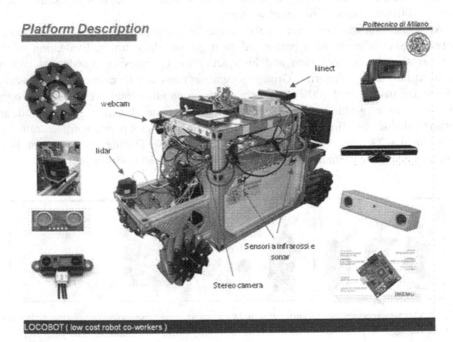

Fig. 5. LOCOBOT Platform

is to precisely position with an accuracy below 2 cm, using a webcam facing the ceiling of the working stations. The use of the double layer architecture enables the robot to perform basic navigation and obstacle avoidance very quickly and without excessive computational load.

One of the tasks of LOCOBOT is to carry loads by its robotic arm. In order to stabilize while stationary and protruding its arm, it was provided with four stabilizer robotic legs controlled by force control loop and actuated by four electric linear actuators, each one with a load cell to measure the vertical load.

4.4   Modular Lightweight Robotic Arm

To be open for future different scenarios, the design of the robotic arm was kept basically modular and easy to adapt to different sizes.

From the first requirements of having a compact "transport" position on the platform and high range of motion and wide outreach, a special kinematic was developed, (Figure 6). Further major requirements were: payload up to 20 kg and range of motion from 70 cm to 130 cm (in the plane). Therefore a minimum number of two configurations (a "short" and a "long" one) was decided. Modularity provides the option of additional combination of modules. A special approach was necessary to cope with a payload of 20 kg, still allowing human robot collaboration.

The design of the whole arm was mainly restricted by the necessity to provide intrinsic safety for the human robot collaboration features. To cope with these requirements a compromise was found between speed and force as well as the degrees of freedom. The movement was divided into "dangerous vertical movement", "soft" horizontal and small "gravity compensated" vertical movement. The robot is composed by a stiff vertical axis for large vertical movements, "soft" actuators for the horizontal part of the trajectory and an active contact flange for gravity compensated small vertical movements. The single rotary modules of LOCOBOT use an internal physical compliant module to become "soft". At the same time, if some unexpected force is applied, the soft actuators also sense this force, and so recognize the danger of collisions or obstacles. So if the arm moves horizontally and collides with an obstacle, the yielding robot cushions the contact and so only a small force is acting on it (due to the physical compliance).

A feedback about the contact situation is given to the path planning system simultaneously, to inform about the undesired behavior. The motion does not necessarily stop in such a case. The robot can also be instructed to wait a certain time, still applying some force. As soon as the inference is solved, the arm can continue its original movement. The same applies for vertical movements. Here the system takes advantage of the compliance of the installed active contact flange (ACF). The ACF compensates the payload, holding it floating. Again in case of collision the force between object and robot is held low, and a signal is sent to the path planner. This force feedback system allows also to move the robot manually in a very easy way. Due to the real-time control of forces exerted on the robot, a human can easily guide TCP to teach program positions (show-do programming).

Fig. 6. Lightweight modular rotary actuators

For the vertical movement a spindle axis was chosen with an integrated guiding rail. The sizing of the axis was done considering the starter and the battery lifting task and of the mass load of the arm itself. For these tasks two different flanges have been designed as a connection between the vertical axis and the rotational drives. To grip down to the bottom of the starter box, a pneumatic two position axis was added between the gripper and pneumatic active contact flange.

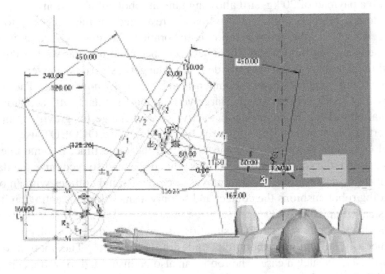

Fig. 7. The payload (yellow rectangle) is moved from the mobile platform (grey square) to the engine bay (blue rectangle), at a distance of approx 700mm from the platform. The robot arm (yellow) is unfolded under the guidance of a person, who stands at the side of the platform.

The high load has to be lifted by a motor driven with 24 V due to the battery of the mobile platform. A new designed integrated drive with a power stage and a controller unit was developed for this application, a holding brake and a gear box is also included. As the controller has a floating point unit it is prepared to implement controllers directly out of Matlab/Simulink models so that a compliant controller could be implemented. If the arm hits an obstacle it becomes soft and the compliance is adjustable. Parallel to the holding brake and motor measurement system in the drive a safety concept with an additional brake and external measurement system is proposed.

Finally a special developed "self adapting" gripper is used for tasks with high variance between parts to be picked. This property is used in the "starter" scenario. The starters are geometrically similar, but slightly different in weight and dimension. To overcome misinterpretations or continuous gripper changes, the flexible "Finray" gripper was designed to cover a wide spectrum of parts. The fingers of this gripper adapt mechanically to different curved surfaces. Therefore it is especially appropriate for cylindrical objects as the starters.

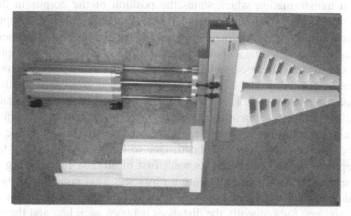

Fig. 8. Finray gripper is self-adaptive to different work pieces, with telescopic cover to prevent injuries

4.5   Safety

An aspect of most importance regarding robotic co-workers is safety. As risk analysis revealed, the major concerns are mechanical hazards caused by collisions. The resulting injuries can be limited by lightweight/compliant mechanical design of the manipulator and post-collision reaction strategies [10]. However lightweight robots have payloads limits, constraining the tasks they can do. Furthermore, even if the injury resulting from a collision is bull or only minor, the acceptance of the system by the human workers is likely to decrease drastically. Thus, safe operation of the robot must be achieved by preventing undesirable contact between robot and human.

The LOCOBOT approach to avoid collisions uses multiple Photonic Mixer Devices (PMD) to monitor the working envelop of the mobile robot. Based on the PMD data a virtual representation of the robot and its current environment is computed, allowing distance calculations between robot and obstacles. Using these distances and motion data of the robot, collision probabilities are predicted in real-time, which are in turn used to control the robot and thus avoid collisions. In contrast to other PMD-based collision avoidance approaches [11], in LOCOBOT a set of environment-mounted sensors is insufficient for the problem due to the mobility of the platform. As consequence of the robot-mounted moving sensors, a more complex pre-processing is required, exacerbating the real-time problem.

Central aspects are the modelling of the robot, the calibration of the sensors, the sensor data pre-processing and the distance computation. This requires modelling of the physical shape of the LOCOBOT system as well as of its kinematics using the Unified Robot Description Format (URDF). Regarding the sensor calibration procedure an Iterative Closest Point (ICP) algorithm was applied, using a calibration corpus with known dimensions mounted on the robots Tool Center Point (TCP). The algorithm registers the observed point cloud of the corpus into a sampled model of it, resulting in a transformation which states the position of the corpus in the sensor coordinate system. This position, via the robot kinematics allows computing the sensor position in the robot coordinate system. The pre-processing of the sensor data has two phases: data filtering and data transformation. In the filtering phase data distorted by sensor-typical errors are identified and abolished. Applied techniques are intensity thresholds, radius or statistical outlier removal strategies. In the second phase the data of different sensors are transformed into a common coordinate system, allowing their interpretation for distance calculation. In case of robot arm mounted sensors the current robot pose has to be incorporated. In the final step the distances between robot and obstacles are calculated. Due to the fact that parts of the robots shape are captured in the sensor data as well, first of all data representing the robot have to be separated from data representing obstacles. This is achieved by aligning the sensor data with a bounding box representation of the current robot pose and removing the inliers. Subsequently the distances between each box and the remaining obstacles are calculated. Having the obstacle distances and positions for each box, an adaption of the robot motion regarding approaching and departing obstacles along the robot structure can be performed.

The described approach has been realized in a prototypical manner for a LOCOBOT prototype and a six degree of freedom industrial manipulator. In first test the sensors have been mounted to the environment as well as to the robot arm, ensuring a sufficient coverage of the working envelop. Using static obstacles and a TCP velocity of 0.65 m/s collisions could be prevented. While these tests demonstrated the feasibility of the approach, they lack of expressiveness regarding its limits. In order to evaluate the systems reactivity, the sensors mounted to the environment will be removed in a next setup, reducing the covered area to a limited section around the manipulator. Static obstacles in the working envelop thus become visible only when they enter the arm-mounted sensors field of view due to the robot motion. Knowing the borders of the sensors field of view and moving the TCP with a defined velocity towards the obstacle, allows thus to calculate the systems reaction time by measuring the distance covered by the TCP after crossing the border. As the evaluation is going on, robust results regarding the reactivity are not yet available. The feasibility of the approach however could already be demonstrated.

The next steps after the first evaluations are concerned with the so far identified weaknesses of the approach, which are the occlusion problem, the workspace coverage and erroneous sensor data. The intended tools to solve these issues are a sensor placement algorithm and a cell-based data representation. The sensor placement algorithm will take into account knowledge about the robots environment and task as well as the field of view of the sensor. Using an offline simulation it will apply this information to compute a sensor positions resulting in minimum object occlusion while covering the relevant workspace. The cell-based data representation will be used as instance for sensor data fusion, similar to [10]. Thereby it allows the identification of remaining occlusions due to its defined spatial structure. Knowing occluded areas enables the robot control to avoid these areas or traverse them with appropriate velocity. On the other hand sensor data can be grouped based on their subsuming cell, allowing a more robust handling of erroneous data.

5   Conclusion

The platform is a highly flexible solution to many problems that now affect the industrial production. For its practical application the industrial requirement of being 100% failure safe with regards to any task performance requires further research and improvement of the different controls and of the HRI modules, on the other side safety controls appears to be satisfactorily implemented.

Acknowledgments. Research supported by The EU Commission FP7 Grant 260101 LOCOBOT (http://www.locobot.eu/).

References

1. Video of the LOCOBOT starters scenario retrievable at,
   http://youtube/i9Fagzy1cxc
2. Barattini, P., Morand, C., Robertson, N.M.: A Proposed Gesture Set for the Control of Industrial Collaborative Robots. In: 21st IEEE International Symposium on Robot and Human Interactive Communication, Ro-Man 2012, Paris, France, September 9-13, p.1.11 (2012)
3. Plasch, M., Pichler, H., Bauer, H., Rooker, M., Ebenhofer, G.: A Plug & Produce Approach to Design Robot Assistants in a Sustainable Manufacturing Environment. In: 22nd International Conference on Flexible Automation and Intelligent Manufacturing, FAIM 2012, Helsinki, Finland, June 10-13 (2012)
4. Video of the robot platform retrievable at,
   http://www.youtube.com/watch?v=F0xEy6qaNoI
5. Wolfel, M., McDonough, J.: Distant Speech Recognition. Wiley, New York (2008)
6. Tashev, I.J.: Sound Capture and Processing: Practical Approaches. Wiley (2009)
7. Young, S.: The ATK Real-Time API for ATK,
   http://htk.eng.cam.ac.uk/develop/atk.shtml
   (retrieved September 14, 2012)

8. Tashev, I.J.: Audio for Kinect: pushing it to the limit (invited talk). In: CREST Symposium on Human-Harmonized Information Technology, University of Kyoto (April 2012)
9. ChaLearn Gesture Challenge, http://gesture.chalearn.org/
10. Luca, A.D., Albu-Schaffer, A., Haddadin, S., Hirzinger, G.: Collision Detection and Safe Reaction with the DLR-III Lightweight Manipulator Arm. In: IEEE/RSJ International conference on Intelligent Robots and Systems, pp. 1623–1630 (October 2006)
11. Flacco, F., de Luca, A.: Multiple depth/presence sensors: Integration and optimal placement for human/robot coexistence. In: IEEE International Conference on Robotics and Automation, ICRA (2010)

Communication Infrastructure in the Centralized Management System for Intelligent Warehouses

Kelen Cristiane Teixeira Vivaldini[1], Gabriel Tamashiro[1], José Martins Junior[2], and Marcelo Becker[1]

[1] EESC-USP, Mechatronics Group - Mobile Robotics Lab., Av. Trabalhador Sao Carlense, 400. São Carlos-SP, Brazil
[2] EEP-FUMEP, Computer Science Dept, Av Monsenhor Martinho Salgot, 560, Piracicaba-SP, Brazil
{kteixeira,becker}@sc.usp.br, gabriel.tamashiro@usp.br, martinsjr@gmail.com

Abstract. The automation of logistic systems is essential to improve productivity and reduce costs. An effective way to introduce automation in materials handling is applying Automated Guided Vehicles (AGVs). The design of an AGV System requires decision making regarding the best strategies to solve the various problems associated with its functioning (efficient routes, AGVs selection, collision, deadlocks, battery management, etc.). In this context, this paper presents a centralized architecture AGV System for intelligent warehouses. In this architecture, the Intelligent Supervisor controls the AGVs and exchanges information with the Warehouse Management System (WMS). A middleware-based communication infrastructure was developed to ensure the communication interoperability among the components of the system and the information manager. By measuring the response time and performing tests of the network overload for a different number of AGVs controlled by the Intelligent Supervisor, a reliable standard to exchange information could be achieved.

Keywords: Communication Infrastructure, AGV system, middleware, intelligent warehouse.

1 Introduction

Materials handling is an essential aspect of any production system and an extremely important aim is the maximizing of the system flexibility. Le-Ahn [1] affirms that in practice, the materials handling substantially contributes to the value of product. Due to the high operational costs attributed to materials handling activities, organizations have been seeking for ways to minimize the time spent on their execution and optimize their operations. According to Beker, Jevtić, and Dobrilović [2] a productivity improvement in the reduction of any costs, especially by eliminating the biggest waste in inner transport, can be obtained by the automation of logistic processes.

P. Neto and A.P. Moreira (Eds.): WRSM 2013, CCIS 371, pp. 127–136, 2013.

Automated logistic systems of distribution areas, such as industries, warehouses, cross docking centers and container terminals frequently use Automated Guided Vehicles (AGVs) to optimize the materials handling tasks [3]. AGVs increase efficiency and reduce costs in a wide range of applications, including distribution logistics [4]. The system for the control of AGVs is referred to as Automated Guided Vehicle System (AGVS) and consists of several AGVs that operate concurrently (including automatic operations, reduced labor and increased productivity) and automated interfaces with other systems.

The design of an AGV System requires decision making regarding the best strategies to solve the several problems associated with its functioning. Based on the requirements imposed by the materials handling system, of which the AGV system is an integral part, it is necessary to choose the solutions that best adapt to the existing requirements.

The main functionalities of AGVS are the insurance of an efficient flow of materials in a warehouse, including flexibility to adapt to changes in requests, assignment transport tasks to appropriate AGVs (estimating the number of vehicles required), efficient routing of the AGVs through the warehouse avoiding collisions and deadlocks, and maintenance of the AGVs' batteries [5-7].

The AGVS in a warehouse can be controlled by two types of architecture, centralized and decentralized [8-11][7][12-16]. The main difference between them is that in a decentralized system, a vehicle operates as an independent agent based on local information, depending on an organized behavior of autonomous vehicles that collaborate to achieve assigned tasks. And, the decentralized architecture aims to solve the coordination only when conflicts arise [12]. The system's performance may be affected by the communication links between nodes. In a centralized control system, a central system controller is responsible for dispatching vehicles by using available information from all possible sources, keeping track of all movements in relation to the handling of materials. The centralized architecture aims to solve a problem considering the whole system, which enables to find the global optimal solution [12]. The centralized controller continuously communicates with vehicles to guide them, providing an easy diagnosis of errors [1][17-19]. Thus, to ensure functionality to the system, the information must be received and sent correctly between WMS and AGVs, because the interaction between the systems is wholly dependent on the type of communication adopted.

This paper presents an AGVS that uses a centralized architecture and takes into account the interaction between the systems to be monitored (AGVs) and the interactions between information exchanged with the WMS (Warehouse Management System). To ensure that the result of the information exchange between systems is correct and at runtime, a middleware-based communication infrastructure was developed to establish and synchronize information and transfer the data at runtime.

1.1   AGVS Architecture

Fig. 1 shows the proposed architecture for the AGVS. In this architecture, the Intelligent Supervisor receives orders from the WMS and operates on AGVs to ensure the successful execution of the work plan. Each order involves multiple tasks, and each task represents the loading, transportation and unloading of a pallet. According to the tasks execution, the WMS receives information about the status of the task.

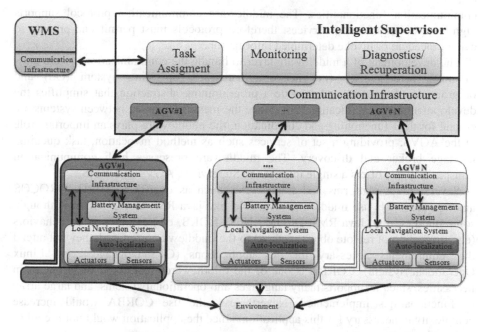

Fig. 1. Architecture proposed for the AGVS

As can be seen in Fig.1, the Supervisor monitors the evolution of the process, action by action. For each order, the Supervisor allocates the tasks selecting the amount of AGVS necessary to execute them and calculates their respective conflict-free routes and deadlocks, optimizing the amount of maneuvers. Then, it sends routes to each AGV and regularly receives information about the execution tasks. Each AGV has its own sensors, actuators, auto-localization, and local navigation subsystem. Taking into account the global route provided by the Supervisor, each AGV calculates its own local path necessary to execute its tasks and monitors its surroundings looking for mobile or unexpected obstacles during the execution of the planned path. The AGVs provide the state information (e.g., sensor readings, actual position, task status and battery data) for the Supervisor, which evaluates the execution status step by step. In case of detection of any deviation, the Supervisor can act before an error occurs avoiding delivery delays, collisions or deadlocks. During the execution of tasks by the AGV, in each loading and unloading of the pallet, a message is sent to the WMS to update the status of the order. Therefore, the communication system is essential for the result of the implementation of the systems to be correct and at runtime, because the interaction between the systems is totally dependent on the type of communication adopted. Therefore, we must ensure a reliable standard to send and receive this information.

2   Communication Infrastructure

The main goal of the middleware abstraction layer is to simplify the development of applications in distributed systems, providing abstraction of the necessary communication functionalities. The middleware communication protocol supports high-level communication services, therefore protocols must permit one process to call a procedure or invoke determined information.

The development of a middleware layer to handle the communication system aims to mask heterogeneity (i.e. network, hardware, operating system (OS) and programming language) and provide a programming abstraction that simplifies the development of the application [20]. Since the messages passed between systems are crucial for their functioning and coordination, the middleware plays an important role in the AGVS, providing a set of services such as method invocation, task queuing, message broker and discovery. The middleware preserves high communication transparency and offers a single interface to Supervisor, AGVs and WMS.

Several robot platforms and frameworks, such as openTCS, MIRO, OROCOS among others [21] use middleware available like java RMI and CORBA. Although, some middlewares (Java RMI, DCOM, .NET, HORB, etc) exhibit similar behaviors (e.g. invocation of remote object method) to the middleware proposed, they not attend the requirements necessary for this applications (C++ language   and Linux Operational System). CORBA is one of the middlewares that could fit to our main necessities, since it supports many languages and operational systems, and large array of functionalities implemented. Nevertheless, the use CORBA would increase complexity unnecessary for this application, since the application would not use all of its functionalities.

The proposed middleware is based on client-server architecture and uses concepts of object-oriented programming (OOP) and multithreads, providing flexibility to the addition of new functionalities and protocols. The connections were established through a router (IEEE 802.11 standards for wireless networks [22]) and the TCP/IP protocols were used. For the discovery service, an idea similar to the Address Resolution Protocol [23] was used: a message containing an identifier is broadcasted using UDP and each device that recognizes the identifier sends back a response message to establish a connection. Events coordination and task ordering were implemented based on message causality, as described by the Lamport Model [24]. A partial class diagram of the middleware is showed in Fig. 2.

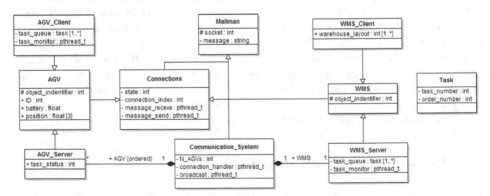

Fig. 2. Partial class diagram of the communication middleware

The TCP/IP and UDP protocols used in the message service were implemented in the Mailman and Connections classes, which are inherited by the WMS, AGV and Communication_sytem classes. As shown in Fig. 2, the WMS_Server and the AGV_Server classes are contained by the Communication_System class and inherit, respectively, WMS and the AGV classes. Similarly, the WMS_Client and the AGV_Client inherit, respectively, the WMS and the AGV classes. The Task class was used in the task queuing service and is contained by the AGV_Client and the WMS_Server classes.

Basically, the whole communication is represented by messages passed among objects of three classes: WMS_Client, AGV_Client and Communication_System. Each object of the WMS_Client and AGV_Client classes represents a single process that to communicates with the supervisor and is implemented in the WMS and AGVs routines, respectively. The object of the Communication_System class implemented in the supervisor routine searches for the WMS and the AGVs connected to the network and associates each of them with an object (WMS_Server and AGV_Server) establishing a connection. Each object spawns an appropriate number of threads that manage the connections and the exchange of messages. Regarding the messages format, besides the TCP/IP header, messages are encapsulated in an additional header used to specify the requested service (see Fig. 3). Fig. 4 illustrates the flow of messages between the Supervisor and WMS connections (A) and between the Supervisor and AGV connections (B).

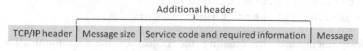

Fig. 3. Message encapsulation

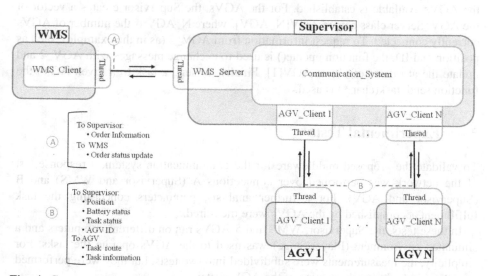

Fig. 4. Communication diagram of the middleware. Some usual messages passed between connection A and B are also presented.

Once the connections have been established (A and B, indicated in Fig.4), the Supervisor interact with the WMS and the AGVs by requesting information and services. In a typical routine, the Supervisor receives order information from the WMS, processes it (generating routes, etc) and sends tasks to the AGVs. All the task data are stored by the middleware queuing service, reducing the applications effort necessary to manage information.

To exemplify the operation of how the communication mechanism, the Supervisor code that establishing the connections, request information from an AGV and sends a given task to this AGV, is presented below.

Example of the Supervisor code that establishes the connections with the AGVs, requests information from a specific AGV and sends a given task to this AGV

```
Char *task;
...
Communication_System Supervisor;
int N = Supervisor.get_N_AGV();
printf("Number of AGVs connected: %d\n", N);
Supervisor->AGV[1].update();
printf("Position of AGV_2 (ID: %d): (%f,%f) (x,y)\n",
Supervisor->AGV[1].ID,
Supervisor->AGV[1].position[0],
Supervisor->AGV[1].position[1]);
Supervisor->AGV[1].send_task(task);
```

As previously mentioned, by creating the object of the Communication_System (named Supervisor) class, the discovery service is triggered and the connection with the AGVs available is established. For the AGVs, the Supervisor creates a vector of the AGV_Server class named AGV[N_AGV], where N_AGV is the number of AGVs currently connected. To request information from AGV_2 (as in the example), such as position and ID, the function update() is used to exchange messages with AGV_2 and update the attributes of object AGV[1]. Finally, to send a task to a given AGV, the function send_task(char *) is used.

3   Experimental Tests

To validate the proposed middleware for the communication system, a response test of the network was performed over connections A (Supervisor and WMS) and B (Supervisor and AGV). For a further analysis, parameters concerning the task fulfillment accomplished by the AGVs were measured.

In these tests, the Supervisor, WMS and 5 AGVs ran on different computers and a simulated environment (Player/Stage) was used to the AGVs operate the tasks. For simplicity, the measurements were subdivided into two tests, but they were performed simultaneously during the routine of the AGV system.

In test 1, the time spent to send and receive one packet of a given size through each connection (A and B) is measured, thus providing the response time of the network. To perform this time measurement during the applications routine, a middleware functionality was implemented to enable the supervisor to perform the response time test at any time. Then, in test 1, the time acquired by the application is compared with the time provided by the ping command between computers. The packet sizes for the tests were chosen based on the most common packet sizes required by the AGV system, and the ping (Internet Control Message Protocol – Echo Request-Reply) command returns the shortest response time given by the network infrastructure. The mean network response times measured by one AGV and five AGVs connected to the Supervisor are provided in Tables 1 and 2.

Table 1. Test 1 - Response time (ms) obtained by 1 AGV connected to the Supervisor

Supervisor	Packet size (characters)	AGV	WMS
At runtime of the	50	3	4
Application	15000	20	38
ICMP – ping	50	2	4
ECHO REQUEST	15000	12	30

Table 2. Test 1 - Response time (ms) obtained by 5 AGVs connected to the Supervisor

Supervisor	Packet size (characters)	AGV 1	AGV 2	AGV 3	AGV 4	AGV 5	WMS
At runtime of the	50	6	5	7	6	6	6
Application	15000	49	31	55	43	57	57
ICMP – ping	50	2	2	4	2	2	6
ECHO REQUEST	15000	33	12	32	37	33	30

Tables 1 and 2 show that for the packet size of 50 characters the mean deviations of the difference between the application and ping are 0.5 ms for one AGV connected and 3 ms for five AGVs. For the packet size of 15000 characters, the differences were 8 ms for one AGV and 19 ms for five AGVs.

As expected, an increase in the number of AGVs connected with the Supervisor also increased the mean response time of the network in the application, but remained in the same order of magnitude to that obtained with the ping command. Therefore, the proposed middleware does not aggregate considerable processing time to degrade the network response time provided by the network infrastructure.

In Test 2, parameters of the order fulfillments executed by the AGVs were measured using one and five AGVs. These parameters provide some time and spatial constraints related to the AGV system, and are shown in Tables 3 and 4.

Table 3. Test 2 - Routines of one AGV connected to the Supervisor.

	AGV_1
Number of order attended	1
Quantity of pallets	27
Mean distance traveled (m)	1193
Execution time of the task (min)	22,80

For one AGV connected, the time required for the Supervisor to generate the routes for the AGV was 163 ms and the time waited by the WMS for the AGVs to start the order fulfillment after it has sent the order to the Supervisor was 181 ms.

Table 4. Test 2 - Routines of five AGVs connected to the Supervisor

	AGV_ 1	AGV_ 2	AGV_ 3	AGV_ 4	AGV_ 5
Number of order attended	11	11	11	11	11
Quantity of pallets	53	50	47	50	46
Distance traveled (m)	8327	8604	6355	8342	7682
Execution time of all task (min)	68,25	55,25	48,48	61,83	62,89

For five AGVs connected, the time required for the Supervisor to generate all routes for the AGVs was 2061 ms and the time waited by the WMS for the AGVs to start the orders fulfillment after it has sent the order to the Supervisor was 197 ms. The difference between these time values is explained by the fact that the Supervisor dispatches the routes to the AGVs as they are generated.

After the start of the order fulfillment, the information is exchanged between the Supervisor and the AGVs for each loading and unloading of the pallets. This information is necessary to update the order status in the WMS. The average time for this message exchange is 25 s. Other messages, such as number of AGVs, task status, current position and battery charge can be requested by the Supervisor and the AGVs can also send messages to the Supervisor in case of unexpected problems.

The time parameters measured in test 2 are at least ten times higher than the mean response time measured in test 1, therefore, the middleware response time is not crucial to the processing time of the application.

4   Conclusions

This paper has described a centralized architecture to manage an AGV System for intelligent warehouses, where the Intelligent Supervisor controls the AGVs and exchanges information about orders with the WMS. A middleware infrastructure was developed to ensure these information exchanges and interoperability between the systems.

The communication infrastructure was validated by two tests that measured the response times of the network and compared them with time constraints relative to the AGV system for different number of AGVs connected. During the tests, the middleware maintained the connections and provided a response time in accordance with the time constraints of the application. The middleware functionalities, such as the discovery service and queuing service also provided a relevant programming abstraction that simplified the application development. Therefore, the proposed communication system reached a reliable standard to exchange messages, implying no limitation on the application. The centralized architecture adopted and the communication infrastructure met all requirements imposed for the correct functioning of the AGVs and enabled the integration of information to update the status of order fulfillment in the WMS.

For future studies, persistence and security related to the network will be implemented in the communication infrastructure. Also, interaction of the AGV system with external databases will be studied to increase its applicability for industries. For the AGV system, functionalities associated with the supervisor, such as forecast of future problems and priority changes of order fulfillment, will also be implemented.

References

1. Le-anh, T.: Intelligent control of vehicle-based internal transport systems, Ph.D Dissertation, ERIM Ph.D series research in management 51, Erasmus University Rotterdam (2005)
2. Beker, I., Jevtić, V., Dobrilović, D.: Shortest-path algorithms as a tools for inner transportation optimization. Int. J. Ind. Eng. and Management 3, 39–45 (2012)
3. Vivaldini, K.C.T., et al.: Automatic Routing System for Intelligent Warehouses. In: IEEE Int. Conference on Robotics And Automation, pp. 93–98 (2010)
4. Klaas, A., et al.: Simulation aided, knowledge based routing for agvs in a distribution warehouse. In: Winter Simulation Conf., Phoenix, AZ, pp. 1668–1679 (2011)
5. Le-Ahn, T., De koster, M.B.M.: A review of design and control of automated guided vehicle systems. European Journal of Operational Research, 1–23 (2006)
6. Vis, I.F.A.: Survey of research in the design and control of automated guided vehicle systems. Eur. J. Op. Research 170(3), 677–709 (2006)
7. Weyns, D., Schelfthout, K., Holvoet, T.: Architecture-Centric Development of an AGV Transportation System. In: Pĕchouček, M., Petta, P., Varga, L.Z. (eds.) CEEMAS 2005. LNCS (LNAI), vol. 3690, pp. 640–644. Springer, Heidelberg (2005)
8. Manca, S., Fagiolini, A., Pallotino, L.: Decentralized Coordination System for Multiple AGVs in a Structured Environment. In: 18th Int. Federation of Automatic Control, pp. 6005–6010 (2011)
9. Singh, N., Sarngadharan, P.V., Pal, P.K.: AGV scheduling for automated material distribution: a case study. J. Intell. Manuf. 22, 219–228 (2011)
10. Fauadi, M.H.F.B.M., Li, W., Murat, T.: Combinatorial Auction Method for Decentralized Task Assignment of Multiple-Loading Capacity AGV Based on Intelligent Agent Architecture. In: Innovations in Bio-inspired Computing and Appl., pp. 207–211 (2011)

11. Pallottino, L., Scordio, V.G., Frazzoli, E., Bicchi, A.: Decentralized cooperative policy for conflict resolution in multi-vehicle systems. IEEE Transaction on Robotics 23(6), 1170–1183 (2007)
12. Pérez, D.H., Barberá, H.M.: Decentralized coordination of autonomous AGVs in flexible manufacturing systems. In: 2008 IEEE/RSJ Int. Conf. Intell. Robots and Systems, pp. 3674–3679. IEEE Press, New York (2008)
13. LaValle, S.M.: Planning Algorithms. Cambridge University Press, Cambridge (2006)
14. Alami, R., Fleury, S., Herrb, M., Ingrand, F., Robert, F.: Multi-robot cooperation in the martha project. IEEE Robotics Automation Magazine 5(1), 36–47 (1998)
15. Lygeros, J., Godbole, D., Sastry, S.: Verified hybrid controllers for automated vehicles. IEEE Transactions on Automatic Control 43(4), 522–539 (1998)
16. Lindgren, H.: Centralised and decentralised control for AGVS: advantages and disadvantages. In: 3rd Int. Conf. on Automated Guided Vehicle Systems, Stockholm, Swed, pp. 209–218 (1985)
17. De Wolf, T., Samaey, G., Holvoet, T., Roose, D.: Decentralised Autonomic Computing: Analysing Self-Organising Emergent Behaviour using Advanced Numerical Methods. In: Autonomic Computing, ICAC 2005, New York, pp. 52–63 (2005)
18. Weyns, D., Schelfthout, K., Holvoet, T., Lefever, T.: Decentralized Control of E'GV Transportation Systems. In: 5th Int. Conf. on Autonomous Agents and Multiagent Systems, Utrecht, Netherlands, pp. 67–74 (2005)
19. Ong, L.: An investigation of an agent-based scheduling in decentralised manufacturing control. Ph.D Disseration, University of Cambridge (2003)
20. Coulouris, G., Dollimore, J., Kindberg, T.: Distributed Systems Concepts and Design. Addison-Wesley, Boston (2011)
21. Iñigo-Blasco, P., et al.: Robotics software frameworks for multi-agent robotic systems development. Robot. Auton. Syst. 60(6), 803–821 (2012)
22. IEEE (Institute of Electrical and Electronic Engineers) 802.11. Local and Metropolitan Area Networks – Part 11: Wireless LAN Medium Access Control (MAC) and Physical Layer (PHY) Specifications (1999)
23. Tanenbaum, A.S., Van Steen, M.: Distributed Systems: Principles and Paradigms. Pearson Prentice Hall, Upper Saddle River (2007)
24. Lamport, L.: Time, clocks and the ordering of events in a distributed system. Comms. ACM 21(7), 558–565 (1978)

Intelligent Warehouse Product Position Optimization by Applying a Multi-criteria Tool

Livia Martinelli Tinelli, Kelen Cristiane Teixeira Vivaldini, and Marcelo Becker

EESC-USP, Mechatronics Group - Mobile Robotics Lab., Av. Trabalhador Sao Carlense,
400. São Carlos-SP, Brazil
{tinelli,kteixeira,becker}@sc.usp.br

Abstract. The operational optimization process applied to warehouses can lead to cost reductions by means of the correct positioning of the products. The definition of the product localization will determine the distance to be traveled to handle the materials and directly impact on warehousing costs. In this context, this article focuses on the positioning optimization of finished products in intelligent warehouses, which allows reducing the distance traveled and the order fulfillment time. The Analytic Hierarchy Process (AHP) was used for the optimization process. This technique consists in hierarchizing the products analyzed by employing pairwise comparisons through the judgment of their characteristics (size, fragility and weight). A simulated intelligent warehouse environment was used to analyze the results of allocation of the product based on AHP tool, where task were distributed for AGVs (Automated Guided Vehicle). As a result, a better positioning of the products and a reduction in the operation time of material handling were achieved.

Keywords: AHP, AGV, seasonal demand, intelligent warehouses optimization.

1 Introduction

The materials handling related to internal logistics occurs (within warehouses) in the operations of input flow, storage and output of products. These operations demand time, labor and money. To avoid unnecessary movements it is necessary to minimize them, and consequently reduce the time of order fulfillment. Bodin et al. [1] show that the physical distribution of products reduces approximately 16% of the final cost of the product. Thus, the correct positioning of finished products contributes to the optimization of costs in the materials handling [2-4], as well as costs related to the maintenance of handling equipment. The location of the finished products in the warehouse will influence the distance traveled to reach the shipping area. The correct positioning of heavier or more fragile products closer to the shipping area and an adequate materials handling in warehouse will result in a cost reduction.

In this context, the Storage Location Assignment Problem (SLAP) determines the area/shelves to store the incoming product items, so that the total operational cost is minimized, reducing material/handling costs and improving the space utilization [5][6]. The SLAP can be classified according to the amount of information known

P. Neto and A.P. Moreira (Eds.): WRSM 2013, CCIS 371, pp. 137–145, 2013.

about the arrival and departure of the products stored in the warehouse [7]. The information and criteria necessary for the SLAP in the definition of the physical arrangement are related to several factors, such as warehouse plan, area/capacity of warehousing, ratio of total positions from the shelves, as well as data on the products to be stored and order processing time.

Hackman and Rosenblatt [8] developed a knapsack heurist procedure to decide on the items and their quantities to be assigned to the forward area. The objective was to minimize the total materials handling costs of order picking and replenishing, providing sufficient conditions for optimality.

Frazelle et al. [9] extended the problem and solution for the method presented [8]. They treated the size of the forward area as a decision variable. Their model includes the equipment cost of the fast pick area (modeled as a linear function of its size) and the materials handling cost for order picking and replenishment.

Frazelle and Sharp [10] and Frazelle [11] describe assignment strategies to determine the locations of components in a warehouse where the total picking tour time is minimized. For Brynzér [12], the general idea is that components likely to appear together in an order should be stored together (dependent demand).

Gu, Goetschalckx and Mcginnis [5] proposed an optimal approach to determine the zones, assignment of items to zones, and base locations in order to minimize the total order picking cost expected. Based on these data, Onüt et al. [13] developed a heuristic based on criteria to sort the items into classes reducing the distances in warehouses and enabling a wide application for SLAP when there are qualitative factors involved in the decision-making process.

In this context, we propose a solution to optimally allocate finished goods in warehouses reducing unnecessary travel costs of the materials handling equipment and maintenance. The SLAP based on product information (PI) was used and the multi-criteria analysis tool, Analytic Hierarchy Process (AHP), was applied to hierarchy the products.

The AHP establishes the criteria (demand, fragility, weight, financial transactions, seasonality, etc.) to be prioritized to allocate products. The best allocation decision is made based on criteria established as a priority, so that products representing a larger share of depiction will be classified in a decreasing order. Thus, the finished products that are better ranked will be allocated next to the shipping area.

2   Analytic Hierarchy Process

Analytic Hierarchy Process (AHP) is a decision-making technique developed by Saaty 1977 [14]. It is a method of multi-criteria decision in complex environments where many variables or criteria are considered in the prioritization and selection of alternatives. It is designed to cope with both rational and intuitive information to select the best alternative from a set of alternatives evaluated with respect to several criteria [15].

According to ODPM [16], AHP, as a compensatory method, assumes complete aggregation among criteria and develops a linear additive model. In this technique, the decision maker performs a simple pairwise comparison judgment of the products characteristics, which are then used to develop overall priorities for ranking alternatives. The weights and scores are achieved basically by pairwise comparisons between all options.

The AHP can be used to construct a decision-making problem, which is a simple hierarchy consisting of three levels: the overall goal of the decision, the criteria by which the alternatives will be evaluated and the available alternatives (Fig. 1). The decision makers can systematically evaluate the alternatives by making pairwise comparisons for each criterion chosen. These comparisons may use concrete data from the alternatives or human judgments as a way to input subjacent information [17][18].

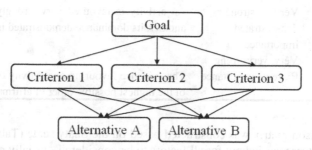

Fig. 1. Saaty Simple AHP Hierarchy [19]

AHP transforms the comparisons, which are most of the times empirical, into numeric values, which are further processed and compared. The weight of each criterion allows the assessment of each element inside the defined hierarchy. According to Vargas [17], this capability of converting empirical data into mathematical models is the main distinctive contribution of the AHP technique in contrast to other comparing techniques. After comparisons have been made and the relative weights between each criterion to be evaluated have been established, the numerical probability of each alternative is calculated. This probability determines the likelihood an alternative will contribute to the expected goal. The higher the probability, the better the chances of the alternative to satisfy the final goal of the portfolio.

AHP is considered a robust method of evaluation that allows the analysis of both quantitative and qualitative variables. The application of the tool is adequate to the context of our study as some qualitative parameters should not be disregarded in the classification of items in warehouses.

2.1   Methodology Adopted

The application of the AHP tool ranks the alternatives, and the highest value of the priority vector indicates the best ranked alternative.

The relative importance scale between two alternatives (Table 1), is the most widely used [20]. As can be seen in Table 1, values from 1 to 9 are attributed to the alternatives and the scale determines the relative importance when compared to another alternative.

Table 1. Fundamental scale of absolute numbers

Intensity of Importance	Definition	Explanation
1	Equal importance	Two activities equally contribute to the objective
2	Weak or slight	
3	Moderate importance	Experience and judgement slightly favour one activity over another
4	Moderate plus	
5	Strong importance	Experience and judgement strongly favour over another, its dominance demonstrated in practice
6	Strong plus	
7	Very strong or demonstrated importance	An activity is favoured very strongly favour over another, its dominance demonstrated in practice
8	Very, very strong	
9	Extreme importance	The evidence favouring one activity over another is of the highest possible order of affirmation

A comparison matrix was constructed based on the Saaty scale (Table 1). Pairwise comparisons were carried out for all factors to be considered - usually not more than 7 - and the matrix was completed. Table 2 shows the scale applied to compare the strategy used for the positioning of finished products. The research was developed in a small company in São Carlos/SP, which provided data about demand, financial transactions and seasonality,

Table 2. Comparison matrix

	Criteria	C1- Demand	C2- Financial transactions	C3-Seasonality
C1	Demand	1	1/7	1/2
C2	Financial transactions	7	1	5
C3	Seasonality	2	1/5	1
	Σ	10	1.343	6.5

According to Vargas [17], it is necessary to normalize the previous comparison matrix by dividing each table value by the total column value (Table 3) to interpret and give relative weights to each criterion.

Table 3. Comparison Normalization Matrix of Criteria

	Criteria	C1- Demand	C2- Financial transactions	C3-Seasonality
C1	Demand	0.1	0.106	0.077
C2	Financial transactions	0.7	0.745	0.769
C3	Seasonality	0.2	0.149	0.154

The contribution of each criterion to the goal is calculated by the priority vector, which shows the relative weights between each criterion (Table 4). The sum of all values from the vector is always equal to one (1) [17].

Table 4. Priority vector Calculation

	Criteria	AHP
C1	Demand	0.094
C2	Financial transactions	0.738
C3	Seasonality	0.168

From the results showed in Table 4, we obtain the most relevant C2 criterion in comparison to the other criteria. To verify if the Consistency Index (*CI*) is adequate, Saaty [21] proposed a Consistency Rate (*CR*), Eq. (1), which is determined by the ratio between the Consistency Index and the Random Consistency Index (*RI*). However, the matrix will be considered consistent if the resulting ratio is lower than 1 [17]. The *RI* value is fixed and based on the number of evaluated criteria, as shown in Table 5.

$$CR = CI/RI < 0.1 \sim 10\%. \tag{1}$$

Eq. 2 provides the new eigenvector $(W)'$ obtained multiplying the comparison matrix (Table 3) and the eigenvector (Table 4).

$$W' = A * W. \tag{2}$$

where: W = Eigenvector
       A = Comparison of the matrix
       W = Eigen vector

According to Saaty [21], the Random Consistency Indices are provided as data of Table 5, where *IR* to $n=3$ is 0.58. The proposed matrix was considered consistent and did not need to be restructured.

Table 5. Table of Random Consistency Indices (*RI*) [21]

n	RI	N	RI	n	RI
1	0.00	6	1.24	11	1.51
2	0.00	7	1.32	12	1.48
3	0.58	8	1.41	13	1.56
4	0.90	9	1.45	14	1.57
5	1.12	10	1.49	15	1.59

The Consistency Rate calculated for the matrix is 0.0116. A Consistency Rate lower than or equal to RC<= 0.10 is considered acceptable, therefore the array is consistent with judgments [21].

The consistency of matrices does not need to be demonstrated because it shows only two alternatives. The new eigenvector is obtained through a joint analysis of the criteria and alternative priority. The value of eigenvector AHP is 0.696 and the eigenvector Random is 0.304. It is evident that the AHP tool criterion has a 69.60% contribution to the goal, whereas the Random allocation criterion contributes with 30.40% to the goal. According to the values obtained, we can validate that the allocation based on the AHP Criterion is better in relation to the random allocation. Section 3 describes the simulation in a warehouse environment to confirm the results.

3   Results

The LabRoM (Mobile Robotics Laboratory) at EESC/USP there are ongoing research using AGVs for material handling. Thus, the same simulated intelligent warehouse environment (40x25m) was adopted so that researches can be integrated.

The positioning of the shelves is fixed, but the areas designated for each hierarchy (high, medium and low) for the allocation were based on the C2 criterion found in Section 2.1 (Fig. 2). The AHP establishes the best allocation decision based on criteria established. Thus, the finished products that are better ranked will be allocated next to the shipping area. The AHP tool was implemented in language C.

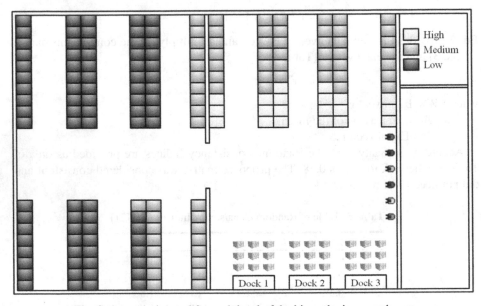

Fig. 2. Areas designated for each level of the hierarchy in a warehouse

To validate the application of the technique the input and output flow of finished products in a warehouse was simulated using loading and unloading tasks of trucks. The Task Assignment Module of Automated Guided Vehicle Systems developed in the LabRoM was used to obtain the distance traveled and the execution time of each task (dock / shelf – unloading task or shelf / dock – loading task). For the Random Allocation and the AHP were used 6 AGVs in the execution of the tasks using a

Player/Stage Simulator. The tests were conducted using 20 tasks (10 unloading and 10 loading of a truck), where each test has 27 pallets.

Table 6 shows the comparison of random and AHP allocations related to the distance traveled and total time of execution of the unloading tasks. As can be seen, a reduction in cost and time could be achieved in all tasks.

Table 6. Comparison between random and AHP allocations for cost and time spent

Unloading Task	Total distance traveled (m)			Total Time of execution (ms)		
	Random	AHP	Reduction (%)	Random	AHP	Reduction (%)
1	618.05	579.80	6.19%	538022	490525	8.83%
2	736.30	585.80	20.44%	632859	480524	24.07%
3	637.30	561.80	11.85%	536855	492525	8.26%
4	652.30	584.30	10.42%	557857	493524	11.53%
5	682.80	547.80	19.77%	587192	476192	18.90%
6	662.05	558.05	15.71%	562356	472025	16.06%
7	682.55	584.30	14.39%	579024	500524	13.56%
8	683.05	598.55	12.37%	577357	521024	9.76%
9	638.55	622.05	2.58%	552690	522690	5.43%
10	631.30	542.30	14.10%	547856	461524	15.76%
Σ	6624.25	5764.7	12.98%	5672068	4911077	13.42%

Table 7 shows presents the same results of the loading tasks.

Table 7. Comparison between random and AHP allocations for cost and time spent

Loading Task	Total distance traveled (m)			Total Time of execution (ms)		
	Random	AHP	Reduction (%)	Random	AHP	Reduction (%)
11	773.70	724.95	6.30%	650790	612292	5.92%
12	665.70	649.45	2.44%	560790	527959	5.85%
13	801.20	711.45	11.20%	668123	588290	11.95%
14	720.45	660.20	8.36%	604289	531122	12.11%
15	715.45	785.45	-9.78%	599957	662625	-10.45%
16	706.20	669.20	5.24%	578789	542124	6.33%
17	724.70	663.30	8.47%	606124	536192	11.54%
18	733.70	731.45	0.31%	619123	600624	2.99%
19	744.45	782.45	-5.10%	617292	661626	-7.18%
20	765.45	673.20	12.05%	633292	544792	13.97%
Σ	7351.00	7051.1	4.08%	6138569	5807646	5.39%

As can be observed in Table 7, some values of the comparison between random and AHP allocations are negative because the routes in the tasks of random allocation were better than those of the AHP. However, the total sum of cost and time was obtained an reduction to the AHP tool.

The analysis shows the total sum presents a reduction of 12.98% to unloading task and 4.08% to loading task related a cost (distance traveled) and a reduction of 13.42% to unloading task and 5.39% loading task related to the total time spent of execution of tasks.

According to the values obtained, we can validate the simulation results that the allocation criterion based on AHP is better with respect to the random allocation.

4   Conclusion

The application of products positioning optimization in intelligent warehouses is focused to minimize costs and labor involved in handling activities. Therefore the optimization of this activity provides a cost reduction. In this context, the Analytic Hierarchy Process (AHP) was used for the optimization process of allocation of the products. Where the products were classified based on criterion established on the AHP, then were allocated in a warehouse. Thus, we presented the main mathematical calculations used for the AHP to provide a proper understanding of the technique and to validate it, the tool was implemented in language C. Some tests were performed and the positioning optimization was achieved using the AHP tool. Significant reductions in the execution time of the tasks and distances traveled were obtained. Therefore, costs spent material handling equipments can be reduced, and consequently maintenance. Thus, the reductions of the times and distances traveled guarantee their activities with lower labor and more agility. In this research were adopted three criteria in AHP, but can be adopt more criteria to make the research more robust. In future works will be applied other tools and the research in a large company.

References

1. Bodin, L.D., et al.: Routing and scheduling of vehicles and crews: the state of the art. Int. J. of Computers and Op. Res. 10(2), 63–211 (1983)
2. Tinelli, L.M., Vivaldini, K.C.T., Becker, M.: Product Positioning using the ABC Method. In: Alfaro, S.C.A., Motta, J.M., Negri, V.J. (eds.) ABCM Symposium Series in Mechatronics, 1st edn., vol. 5, pp. 1–12. ABCM, Rio de Janeiro (2012)
3. Sarker, B.R., Diponegoro, A.: Optimal production plans and shipment schedules in a supply chain system with multiple suppliers and multiple buyers. Eur. J. Op. Res. 194(3), 753–773 (2009)
4. Chen, K.K., Chang, C.T.: A seasonal demand inventory model with variable lead time and resource constraints. Applied Mathematical Modelling 31(11), 2433–2445 (2007)
5. Gu, J., Goetschalckx, M., Mcginnis, L.F.: Research on warehouse operation: A comprehensive review. Eur. J. Op. Res. 177, 1–21 (2007)

6. Xu, J., Lim, A., Shen, C., Li, H.: A Heuristic Method for Online Warehouse Storage Assignment Problem. In: IEEE International Conference on Service Operations and Logistics, and informatics, pp. 1897–1902. IEEE Press, New York (2008)
7. Carlo, H.J., Giraldo, G.E.: Optimizing the rearrangement process in a dedicated warehouse. In: Ellis, K.P., Gue, K.R., De Koster, R.B.M., Meller, R.D., Montreuil, B., Ogle, M.K. (eds.) Progress in Material Handling Research, pp. 39–48 (2010)
8. Hackman, S.T., Rosenblatt, M.J.: Allocating items to an automated storage and retrieval system. IIE Transactions 22(1), 7–14 (1990)
9. Frazelle, E.H., Hackman, S.T., Passy, U., Platzman, L.K.: The forward-reserve problem. In: Ciriani, T.A., Leachman, R.C. (eds.) Optimization in Industry, vol. 2, John Wiley & Sons Ltd., New York (1994)
10. Frazelle, E.H., Sharp, G.P.: Correlated assignment strategy can improve any order-picking operation. Ind. Eng. 21(4), 33–37 (1989)
11. Frazelle, E.H.: Stock Location Assignment and Order Batching Productivity. Ph.D. Thesis, Georgia Institute of Technology, Atlanta. Georgia (1990)
12. Brynzér, H., Johansson, M.I.: Storage location assignment: Using the product structure to reduce order picking times. Int. J. Production Economics 46-47, 595–603 (1996)
13. Onüt, S., Tuzkaya, U.R., Dogac, B.: A particle swarm optimization algorithm for the multiple-level warehouse layout design problem. Computers & Industrial Engineering 54, 783–799 (2008)
14. Saaty, T.L.: A scaling method for priorities in hierarchical structures. J. Mathematical Psychology 15, 234–281 (1977)
15. Saaty, T., Vargas, L.G.: Models, Methods, Concepts and Applications of the Analytic Hierarchy Process. Kluwer, Boston (2001)
16. Office of the Deputy Prime Minister (ODPM, Government UK, 2004). DTLR multi–criteria analysis manual. Corporate Publication (2004), Internet: http://www.communities.gov.uk/index.asp?id=1142251
17. Vargas, R.V.: Using the Analytic Hierarchy Process (AHP) to Select and Prioritize Projects in a Portfolio. In: PMI Global Congress 2010, North American, Washington-DC (2010)
18. Saaty, T.L.: Relative Measurement and its Generalization in Decision Making: Why Pairwise Comparisons are Central in Mathematics for the Measurement of Intangible Factors: The Analytic Hierarchy/Network Process. Madrid: Rev. R. Acad. Cien. Serie A. Mat. 102(2), 251–318 (2008)
19. Saaty, T.L.: How to make a decision: The Analytic Hierarchy Process. Eur. J. Op. Res. 48, 9–26 (1990)
20. Saaty, T.L.: Decision making with the Analytic Hierarchy Process. Int. J. Services Sciences 1(1), 83–98 (2008)
21. Saaty, R.W.: The Analytic Hierarchy Process: What it is and how it is used? Mathematical Modelling, Great Britain 9(3-5), 161–176 (1987)

Recognizing Industrial Manipulated Parts Using the Perfect Match Algorithm

Luís F. Rocha[1,2], Marcos Ferreira[1,2], Germano Veiga[2], A. Paulo Moreira[1,2], and Vítor Santos[3]

[1] Department of Electrical and Computers Enginneering, Faculty of Engeeneering,
University of Porto, Portugal
{luis.andre,marcos.ferreira,amoreira}@fe.up.pt
[2] INESC TEC – INESC Technology and Science (formerly INESC Porto), Portugal
{luis.f.rocha,marcos.a.ferreira,antonio.p.moreira,
germano.veiga}@inescporto.pt
[3] University of Aveiro
vitor@ua.pt

Abstract. The objective of this work is to develop a highly robust 3D part localization and recognition algorithm. This research work is driven by the needs specified by enterprises with small production series that seek for full robotic automation in their production line, which processes a wide range of products and cannot use dedicated identification devices due to technological processes. With the correct classification of the part, the robot will be able to autonomously select the correct program to execute. For this purpose, the Perfect Match algorithm, which is known by its computational efficiency, high precision and robustness, was adapted for object recognition achieving a 99.7% of classification rate. The expected practical implication of this work is contributing to the integration of industrial robots in highly dynamic and specialized lines, reducing the companies' dependency on skilled operators.

Keywords: Industrial Manufacturing, Robotics, Object Recognition, Perfect Match.

1   Introduction

For a long time industrial manufacturing has been taking competitive factors into consideration, such as time, cost and quality. However, modern manufacturing is characterized by customization, which can be accomplished by reducing lot sizes, increasing product variability and short production times. The understanding of these multidimensional challenges leads to the use of techniques and tools which can improve manufacturing processes, as well as decrease and eliminate non-value activities. In this sense, and to maintain their competitiveness in the actual market, industrial manipulators must follow this technology evolution with the penalty of starting to be used only in repetitive processes or mass production strategies. One of their most limiting characteristics accepted as, from a flexible manufacturing point

P. Neto and A.P. Moreira (Eds.): WRSM 2013, CCIS 371, pp. 146–157, 2013.

of view, is their programming procedure. Typically this programming is a fairly time consuming process and represents a high investment, unaffordable for small companies.

However, this is not the only obstacle of industrial manipulators that prevents them from being used in diversified fields of industry. The lack of capacity that they demonstrate in detecting and locating three dimensional objects and also the inflexibility of previously defined motion paths makes it impossible for them to be applied in highly dynamic production environments. These characteristics are at odds with the actual state of the industry, and thus other approaches are required in which the developments verified at the level of sensors and actuators (namely vision and laser systems) open possibilities for designing and developing new solutions particularly through integration. With all this in mind, this paper describes the efforts made in close collaboration with an industrial partner to develop a 3D object recognition and localization system to equip an industrial manipulator performing coating tasks. The system presented can recognize the product in hand and allow the robot to autonomously upload the correct program to execute. Moreover, the product pose information will be sent to the robot and if necessary trajectories adjustments are made. Note that this procedure should be done without having to completely reprogram the industrial robot and without the need for human intervention. Furthermore, the idea is that the developments made and all the architecture presented as part of this work can be extrapolated to other applications.

The aim of the research introduced here arises from a real coating industrial problem presented by FLUPOL. This enterprise is an industrial coating applicator whose goal is to cooperate to solve problems of surface adhesion, dry lubrication or corrosion. Their technological process demands for a very high degree of specialization of their coating operators, as well as a great flexibility of the means of production given the large range of different parts that are treated. Today, FLUPOL's R&D activities focus on the development of a robotized cell that allows a specialized coating operator to directly train the industrial robot. It is also expected that the system will be able to identify the part type that has to be coated.

The production line in this Portuguese enterprise is characterized by a closed conveyor line where the parts are transported vertically and where the coating operations and heat treatment are applied. Furthermore, each part can go through these two operations several times without leaving the conveyor. This production procedure makes it impossible to use other identification systems sensors, such as RFID for part identification. Furthermore, the system's ability to be immune to the different positions of the parts is also seen as a benefit to FLUPOL. In addition, CAD models of the parts are not always available.

This article is organized as follows: Chapter 2 presents a short state of art in the fields of 3D model extraction and object recognition. Chapter 3 presents the laboratory prototype built and some extracted part models stored in a database. Chapter 4 explains the algorithm used in this work for object recognition. The system's final architecture and results are presented in Chapter 5 and 6, respectively, and Chapter 7 presents some conclusions.

2   State of the Art

Having presented the problem, the state of the art can be divided into two different sections: 3D Model Reconstruction Sensors, and Feature Extraction and Object Classification.

2.1   3D Model Reconstruction Sensors

A wide range of sensors are available for 3D modeling. In [1], a reasonable set of three dimensional image reconstruction techniques are presented including, Structured light sensor, Stereo Vision, Photometry, Time of flight. In the field of Structured Light Sensors, [2] presents a research with 2D Laser Range Finders (LRF) to perform 3D scene reconstruction; however, this is usually done in mobile navigation and not with industrial systems. The disadvantages of LRF are their high price for high precision measurements and the measurement variation with the object's respective properties. The laser triangulation systems are the most common non-contact method in industrial equipment such as coordinate measurement machines [3].

Nowadays, Microsoft's Kinect sensor receives most of the attention for 3D modeling [4] due to its low cost. Although the Kinect has a high potential because it is capable of extracting 3D Point clouds adding the color feature, its resolution falls short comparatively to other solutions.

Finally, for the time-of-flight approaches, Yan CuiSchuon et al [5] describe a method for 3D object scanning by aligning depth scans that were taken from around an object using a time-of-flight camera. The authors refer that their approach overcomes the sensor's random noise and the presence of a non-trivial systematic bias, by showing good quality 3D models with a sensor with such low quality data. As previously mentioned, considering that the object in the industrial partner is transported vertically in a low speed (0.01m/s) conveyor, and due to high precision needs, the camera laser triangulation system (structured light) was selected for the application.

2.2   Feature Extraction and Object Classification

3D models contain a significant amount of information that can be analyzed, making it possible to extract fundamental characteristics from the scene. Object recognition is coarsely composed of two steps: feature extraction and object classification. Considering the problem presented the work parts are distinguished only by their shape/pattern and dimensions. Several shape feature extraction techniques are available in the literature, which are carefully surveyed by Yang Mingqiang et al in [6]. Therefore, to discriminate different objects, it is simply necessary to distinguish the parameter/feature value belonging to each class [7]. In the image analysis field and for feature extraction purposes, one of the most used approaches for evaluating object shapes is determining the invariant moments as they do not depend on scaling, translation and rotation [8]. Although that is one of the most well-known approaches, others such as Fourier descriptors, eigenvalues of Dirichlet Laplacian [9] and wavelet descriptors have been developed to describe the shape of different patterns [10]. Having captured unique features resorting to some of the techniques referred before, it is necessary to explore object classification techniques. In this area, the most well-known strategies are those from the fields of pattern recognition and Machine learning (such as k-Nearest Neighbor, Support Vector Machine, Neural Networks, Hidden Markov Models and Bayesian approaches) and Point Cloud Analysis. SVM are a relatively recent approach used for binary classification. Numerous publications are focused on combining feature extraction and machine learning techniques. In [11], a fingerprint matching scheme based on transform features and their comparison is presented. The Discrete Cosine, the Fast Fourier and the Discrete Wavelet Transform were used to extract unique characteristics. Then, the Euclidean distance is used to classify the fingerprint minimum and to compare two feature vectors. The authors claim that the Discrete Cosine and the Fast Fourier presented better results than the Discrete Wavelet Transform, achieving a percentage recognition rate of 87.5 percent. It is worth mentioning that valid research efforts have been made on object recognition that do not rely on Machine Learning, namely using pattern recognition and direct template matching. As an example for recognizing light signals for the Autonomous Driving Competition Robotica 2011, [12] presents a combination of two techniques based on blob analysis and pattern recognition. Their approach consists of applying blob analysis to extract the properties of a pre-segmented image region. Then, in order to perform an adequate detection of signs, a comparison with some reference symbols was used with very high recognition accuracy of symbols for distances to the object up to 2m. This case worked properly because shapes were simple and limited to classes and relatively distinctive among them. Although, all valid approaches high classification rates, flexibility to introduce new parts in the industrial productions line and reliability are all industrial requisites.

3   Laser CCD Camera Triangulation System

As previously mentioned, the camera laser beam triangulation system to create the 3D model for the parts was the approach selected. Measurements were taken to have a structured light environment crucial for capturing images using CCD cameras. This solution was discussed with FLUPOL and no obstacles were raised. Fig. 1 shows the laboratory setup built with FLUPOL to test the system, which allowed the execution of preliminary tests [13]. In the proposed setup, the laser and the CCD Camera (Characteristics: gray image and 1024x768resolution) are located in a central position relatively to the part. The part is then fixed using a support attached to the conveyor that allows the part to move. This makes it possible to produce the required motion for the CCD Camera and the laser beam triangulation system in order to extract the 3D model. With the entire system calibrated, eleven 3D models of different parts were extracted and saved in a local database. Fig. from 2 to 5 presented below show eleven captured 3D models and the corresponding real part. In the same figures, the support where the parts are transported is clearly visible.

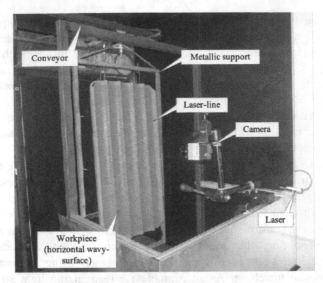

Fig. 1. Laboratorial FLUPOL's set-up

Fig. 2. Part Model – Type A to C (left to right)

Fig. 3. Part Model – Type D to F (left to right)

Fig. 4. Part Model – Type G to I (left to right)

Fig. 5. Part Model – Type J and K (left to right)

4 Pattern Recognition – The Perfect Match

The algorithm for object recognition developed consists of 3D point cloud direct matching. The idea is to compare the 3D model of the part passing in the conveyor (with unknown class) with previously recorded and known class models saved in a database. The matching with the smallest error value will be the class of the unknown part. To perform this matching, the algorithm presented in [14] was used, which has recently been adapted for 3D Matching by Miguel Pinto et al. [15]. High precision, robustness and computational efficiency are some of the reasons which make this algorithm broadly used for mobile robot localization. Therefore, the idea presented in this paper is to extrapolate the use of this matching algorithm to recognise parts for industrial purposes considering FLUPOL's production process.

In [15], the authors started by acquiring a 3D Map of the environment using a laser range finder coupled to the robot. After creating this 3D Map, and considering an offline mode, a distance map and a gradient map were created and stored. The stored distance and gradient matrices are used as look-up tables for the 3D matching localization procedure in the normal robot motion. To create the distance matrix, the distance transform is applied in the 3D occupancy grid of the world map. Furthermore, the Sobel Filter, again in the 3D space, is applied to obtain the gradient matrices, in both the x and y directions. Establishing a parallel with the application presented in this paper, the 3D Map represents the 3D model of each part produced by FLUPOL. Therefore, for each model it is necessary to store equivalent 3D distance data and the gradient matrixes along x (width) and y (height).

Another important parallel has to do with the variables in the localization problem and in the matching problems for parts produced by FLUPOL. Therefore, the objective of the mobile robotic system is to estimate the pose (x, y and θ) in the 3D Map. For the FLUPOL cases beyond part classification, with this approach it will be possible to estimate the displacement x and y and the orientation (along z axis – depth) of the new model when compared to the stored one (state X_m). Note that all the 3D model points are in the world reference frame, defined in the triangulation camera laser calibration procedure.

$$X_m = \begin{bmatrix} x_m & y_m & \theta_m \end{bmatrix}.$$ (1)

In this sense, for the stored models $X_m = [0,0,0]^T$. Now consider a list of laser Camera Triangulation Points (unknown Model) already converted to a data Matrix $[x_i^L, y_i^L, z_i^L]^T$. In this sense, it is possible to write:

$$
\begin{bmatrix} x_i^{Lnew} \\ y_i^{Lnew} \\ z_i^{Lnew} \end{bmatrix} = \begin{bmatrix} x_m \\ y_m \\ 0 \end{bmatrix} + \begin{bmatrix} \cos\theta_m & -\sin\theta_m & 0 \\ \sin\theta_m & \cos\theta_m & 0 \\ 0 & 0 & 1 \end{bmatrix} \times \begin{bmatrix} x_i^L \\ y_i^L \\ z_i^L \end{bmatrix} .
\tag{2}
$$

The 3D Perfect Match [15] runs in the following two steps: 1) Matching error, 2) Optimization routine using Resilient Back-Propagation (RPROP). These two steps are performed until the maximum number of iterations is reached. The distance matrix, stored in memory, is used to compute the matching error. The matching error is computed using the cost value of the list Laser Camera Triangulation points changed $[x_i^{Lnew}, y_i^{Lnew}, z_i^{Lnew}]$:

$$
E = \sum_{i=1}^{N} E_i .
\tag{3}
$$

$$
E_i = 1 - \frac{L_c^2}{L_c^2 + d_i^2} .
\tag{4}
$$

where d_i and E_i are representatives of the distance matrix and the cost function for the laser camera triangulation points $[x_i^{Lnew}, y_i^{Lnew}, z_i^{Lnew}]$. N is the number of Laser Camera Triangulation points. L_c is an adjustable parameter fixed for all experiments. The value tuned for the presented problem is 0,5. This error function was considered in sense to increase its immunity to models outliers.

Computed the matching error the RPROP is applied to each model variable X_m. Therefore the algorithm takes the previous computed state model and uses it in the RPROP iteration. The initial model state is $X_m = [0,0,0]$ since it is considered that the part has the same pose as the corresponding pose stored.

For the RPORP algorithm to execute the distance and gradient matrices (∇x and the ∇y) stored in memory are used for the present part estimated state.

The RPROP routine can be described as follows: during a limited number of iterations, the next steps are performed on each variable to be estimated, x_m, y_m and θ_m.

1) If the actual derivatives $\partial E(t)/\partial x_m$, $\partial E(t)/\partial y_m$ and $\partial E(t)/\partial \theta_m$, depending on the variable, are different from zero, they are compared with the previous derivatives, $\partial E(t-1)/\partial x_m$, $\partial E(t-1)/\partial y_m$ and $\partial E(t-1)/\partial \theta_m$. Where E(t) means E in iteration t.

2) If the product $\partial E(t)/\partial X_m * \partial E(t-1)/\partial X_m$ (for each state variable) is lower than zero, it means that the algorithm already passes a local minimum, and then the direction of the convergence needs to be inverted.

3) If the product $\partial E(t)/\partial X_m * \partial E(t-1)/\partial X_m$ (for each state variable) is higher than zero, it means that the algorithm continues to converge to the local minimum, and then the direction of the convergence should be maintained with the same value.

The limitation of the number of iterations in the RPROP routine makes it possible to guarantee a maximum time of execution for this algorithm

The gradient $\partial E/\partial X_m$ for the state X_m is given by the expression:

$$\frac{\partial E}{\partial X_m} = \sum_{i=1}^{N} \frac{\partial E_i}{\partial X_m}.$$

(5)

$$\frac{\partial E_i}{\partial X_m} = \frac{2L_c^2 \times d_i}{(L_c^2 + d_i^2)^2} \times \frac{\partial d_i}{\partial X_m}.$$

(6)

Where $\partial E_i/\partial X_m$ is the gradient of the cost function of each point i. The partial derivatives, $\partial d_i/\partial X_m = [\partial d_i/\partial x_m; \partial d_i/\partial y_m; \partial d_i/\partial \theta_m]$, are given by the following vector:

$$\frac{\partial d_i}{\partial X_m} = [\frac{\partial d_i}{\partial x_i^{Lnew}} \times \frac{\partial x_i^{Lnew}}{\partial x_m}; \frac{\partial d_i}{\partial y_i^{Lnew}} \times \frac{\partial y_i^{Lnew}}{\partial y_m}; \frac{\partial d_i}{\partial x_i^{Lnew}} \times \frac{\partial x_i^{Lnew}}{\partial x_m} + \frac{\partial d_i}{\partial y_i^{Lnew}} \times \frac{\partial y_i^{Lnew}}{\partial y_m}]$$

(7)

Using the equations presented in (2), the vector (5) can be re-written as the following expressions:

$$\frac{\partial d_i}{\partial X_m} = [\nabla x(P_i); \nabla y(P_i); \begin{bmatrix} \nabla x(P_i) \\ \nabla y(P_i) \end{bmatrix}^T \begin{bmatrix} -\sin\theta_m & -\cos\theta_m \\ \cos\theta_m & -\sin\theta_m \end{bmatrix} \begin{bmatrix} x_i^{Lnew} \\ y_i^{Lnew} \end{bmatrix}].$$

(8)

$$P_i = [x_i^{Lnew}, y_i^{Lnew}, z_i^{Lnew}].$$

(9)

Where $\nabla x[x_i^{Lnew}, x_i^{Lnew}, z_i^{Lnew}]$ and $\nabla y[x_i^{Lnew}, x_i^{Lnew}, z_i^{Lnew}]$ are the gradient values at the position $[x_i^{Lnew}, x_i^{Lnew}, z_i^{Lnew}]$ of the precomputed gradient matrices, stored in memory, in the x and y directions, respectively.

For more information on some details about algorithm please refer to [14].

5   System Architecture

After discussing the matching algorithm, this chapter provides an overview of the system architecture. The system can be divided into two steps: Teaching and Production Phase (see Fig. 6).

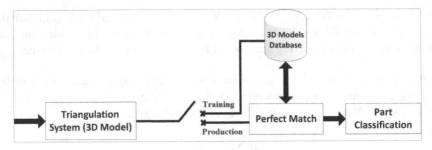

Fig. 6. System Architecture

The teaching phase consists of acquiring a 3D model of the part to be produced in the production line. Therefore, the operator will insert this part in the conveyor and the developed system will capture/store its 3D model and compute the distance map and gradient maps required for the matching algorithm. This procedure only needs to be performed once for each type of part. In the end, a database with the taught parts and respective name will be created dynamically. For the production phase, the operator only needs to insert the already taught part type in the production line. Then, using the Perfect Match algorithm presented before, the model of the unknown part will be compared with the ones in database. For that, the Perfect Match algorithm will be used. Then this classification is communicated to the industrial robot and the correct coating program is uploaded. If the operator inserts a part that was not yet taught, two situations may occur: by evaluating the magnitude of the matching error it is considered that the part is not recognizable or that it is misclassified. Other than identifying the part, the displacement of the part is also computed in comparison with the one in the database. Therefore, and assuming that the industrial robot was taught to perfectly coat the models saved in database, it will be possible to send the trajectory adjustments to the robot along x, y and the rotation along the z axis.

6   Matching Results

After the parts are detected and saved in the database, this section presents the classification of an unknown part (simulation of a production procedure). Basically the unknown 3D model will be compared to the labeled models saved in the database using the Perfect Match Algorithm. The match with the minimum error is the label of this unknown part. The procedure is as follows: firstly, the distance and gradient maps are loaded for a specific type of part saved in the database. Then, by using the Matching Algorithm the matching error (cost value E) is computed, as well as the displacement (Xm) of the unknown model (Figure 6). This routine is performed to all types of parts recorded in database (match the unknown model with all types of parts recorded). In the end the model compared with the least cost value is the correct classification (Table 1 and Figure 7).

Table 1. Matching results for the unkown part with eleven models

Matching Unknown	Cost Value (E)	X coordinate axis correction (meters)	Y coordinate axis correction (meters)	Rotation angle (rad)
vs. Part Type A	1.84	0.0286	0.0017	-0.00028
vs. Part Type B	24.75	0.0143	0.0064	-0.00021
vs. Part Type C	16.57	0.0284	-0.0160	0.02181
vs. Part Type D	8.48	0.0116	0.0039	-0.00161
vs. Part Type E	50.33	0.0702	-0.0714	-0.00420
vs. Part Type F	14.48	0.0009	-0.0005	0.00130
vs. Part Type G	15.35	0.0109	0.0066	0.01516
vs. Part Type H	47.54	0.0163	0.0122	-0.02627
vs. Part Type I	70.72	0.0418	0.0554	0.04115
vs. Part Type J	60.24	-0.0165	-0.0605	-0.01571
vs. Part Type K	40.26	-0.0190	-0.0578	0.02058

Fig. 7. Example of matching between the unknown model and 3 stored ones

In the laboratory setup, the production of 362 parts was simulated belonging to eleven different classes, and a classification rate of 99.7% was achieved. The presence of a great amount of noise in the parts is one of the reasons why the classification is difficult.

6.1 Processing Time

Although good results have been achieved, one of the major problems is related with the processing time. In this sub-chapter, a short study is performed on the number of model points and iterations. The results presented before were made for 100 iterations for RPROP, and using all the points from the data model structure (matrix) to perform the matching. For each matching test the estimated processing time is around 2s (0.5 s to load the distance matrix and gradient x and y matrix and 1.2s to perform the matching). The loading is related to the size of the matrix and to the precision required for the application. A 500 x 300 matrix was considered with a 2.5 mm resolution. The number of iterations of the RPROP is the parameter that controls the computational speed of the Perfect Match. Although one of the important aspects is the processing time, estimating the displacement precision is also a significant task.

Therefore the error was minimized with 300 iterations of RPROP with a computational cost of 4s. This is not satisfactory for the considered purpose. This way, a down sampling of 2 in the 3D model was made, achieving matching times of 530 ms, with a 2% error increase. Summing this matching time with the matrices load time the algorithm computational cost is about 1s (tested performed by using an .Intel Core 2.93 GHz). Although good results have been achieved, one of the major problems is related with the processing time. In this sub-chapter, a short study is performed on the number of model points and iterations. The results presented before were made for 100 iterations for RPROP, and using all the points from the data model structure (matrix) to perform the matching. For each matching test the estimated processing time is around 2s (0.5 s to load the distance matrix and gradient x and y matrix and 1.2s to perform the matching). The loading is related to the size of the matrix and to the precision required for the application. A 500 x 300 matrix was considered with a 2.5 mm resolution. As previously mentioned, the number of iterations of the RPROP is the parameter that controls the computational speed of the Perfect Match. Although one of the important aspects is the processing time, estimating the displacement precision is also a significant task. Therefore, for the application presented here, the error was minimized with 300 iterations of RPROP with a computational cost of 4s.

This is not satisfactory for the considered purpose. This way, a down sampling of 2 in the 3D model was made, achieving matching times of 530 ms, with a 2% error increase. Summing this matching time with the matrices load time the algorithm computational cost is about 1s (tested performed by using an .Intel Core 2.93 GHz).

7   Conclusions

This article presents an algorithm which is robust to noise, reliable in terms of classification and computationally efficient. Furthermore, the algorithm does not depend on any prerequisite of the object, such as the CAD model or known features. For each type of part it is only necessary to store a cloud model in a database, which will be used in the matching algorithm. Another major advantage is that this solution can be extrapolated to others applications where the direct matching between the part and a model captured previously is possible. After earlier results, it was possible to achieve a 99.7 % classification rate. This solution is presently being assembled at FLUPOL where future tests will be performed. Another important aspect is that with the increase of models in the database, although the perfect Match computational time was minimized, the amount of matchings' that will be performed will increase and may reach a processing time that is slower than the one desired for the production line. Therefore the idea is to introduce Support Vector Machine or Neural Networks that are significantly faster computationally, considering the amount of classes, to perform a screening at the beginning of the classification.

Acknowledgments. The work presented in this paper, being part of the Project **PRODUTECH PTI (nº 13851) – New Processes and Innovative Technologies for the Production Technologies Industry**, has been partly funded by the Incentive System for Technology Research and Development in Companies

(SI I&DT), under the Competitive Factors Thematic Operational Programme, of the Portuguese National Strategic Reference Framework, and EU's European Regional Development Fund".

The authors also thanks the FCT (Fundação para a Ciência e Tecnologia) for supporting this work trough the project PTDC/EME-CRO/114595/2009 - High-Level programming for industrial robotic cells: capturing human body motion.

References

1. CustomPacker – Highly Customizable and Flexible Packaging Station for mid- to upper sized Consumer Goods using Industrial Robots, http://www.custompacker.eu (access Date March 2013)
2. World Robotics, Industrial Robots. Published by the IFR Statistical Department, hosted by VDMA Robotics + Automation, Germany (2011) ISBN 978-3-8163-0635-1
3. ABB FlexPicker™, http://www.abb.com/product/seitp327/cf1b0a0847a71711c12573f40037d5cf.aspx (access Date March 2013)
4. Jordt, A., Fugl, A.R., Bodenhagen, L., Willatzen, M., Koch, R., Petersen, H.G., Andersen, K.A., Olsen, M.M., Krüger, N.: An Outline for an Intelligent System Performing Peg-in-Hole Actions with Flexible Objects. In: Jeschke, S., Liu, H., Schilberg, D. (eds.) ICIRA 2011, Part II. LNCS, vol. 7102, pp. 430–441. Springer, Heidelberg (2011)
5. Pichler, A., Ankerl, M.: User Centered Framework for Intuitive Robot Programming. In: International Workshop on Robotics and Sensors Environments (ROSE), Phoenix, Arizona, USA, October 15-16 (2010)
6. Pichler, A., Wögerer, C.: Towards Robot Systems for Small Batch Manufacturing. In: IEEE International Symposium on Assembly and Manufacturing (ISAM 2011), Tampere, Finland, May 25-27 (2011)
7. Festo Corporate – BionicTripod with FinGripper, http://www.festo.com/cms/en_corp/9779.htm (access Date March 2013)
8. Anne, Y.: Balloon filled with ground coffee makes ideal robotic gripper, October 25. Cornell University, Chronicle (2010), http://news.uchicago.edu/article/2010/10/25/balloon-filled-ground-coffee-makes-ideal-robotic-gripper-research-shows
9. Amend, J.R., Brown, E., Rodenberg, M., Jaeger, H., Lipson, H.: A Positive Pressure Universal Gripper Based on the Jamming of Granular Material. IEEE Transaction of Robotics 28, 341–350 (2012)
10. SCHUNK – SDH, http://www.schunk-modular-robotics.com/left-navigation/service-robotics/components/actuators/robotics-hands/sdh.html (access Date March 2013)
11. Dexterous Robot Gripper – Adaptive Gripper, http://robotiq.com/en/adaptive-gripper/ (access Date March 2013)
12. ReconstructMe, http://www.reconstructme.net (access Date March 2013)
13. Lorensen, W.E., Cline, H.E.: Marching Cubes: A high resolution 3D surface construction algorithm. Computer Graphics 21(4) (July 1987)
14. Besl, P.J., McKay, N.D.: A Method for Registration of 3-D Shapes. IEEE Transaction on Pattern Analysis and Machine Intelligence 14, 239–256 (1992)
15. FerRobotics – Robotis with physical compliance. FerRobotics Compliant Robot Technology GmbH, http://www.ferrobotics.at (access Date March 2013)

Flexible Grasping of Electronic Consumer Goods

Martijn Rooker[1], Alfred Angerer[1], Jose Capco[1], Christoph Heindl[1], Aitor Olarra[2],
Elena Fuentes[2], Christian Wögerer[1], and Andreas Pichler[1]

[1] Robotics and Adaptive Systems
PROFACTOR GmbH
Steyr-Gleink, A-4407, Austria
Christian.woegerer@profactor.at
[2] Mechanical Engineering Unit
Teknier
Gipuzkoa, 20600, Spain

Abstract. Automated packaging is becoming more and more interesting for production sites. However, in many companies the packaging process can only handle a limited amount of products. Most of the time for new products, the packaging system needs to be modified. One of the main components in an automated packaging process is the grasping of the objects that need to be packed. For different products mostly a different gripping system is required. In this paper, solutions from a research project are presented which enables the flexible grasping of different products. These solutions include flexible pose recognition for different objects, path planning for the recognized objects and finally, a prototypical design for a flexible gripper, which enables handling of various objects of different weight and size.

Keywords: Flexible packaging, Grasping, Robotics.

1 Introduction

Electronic consumer goods, e.g. dish washers, TV sets, toasters and microwaves have a large number of variants and are packaged manually at most production sites. Only in single-variant production lines with high lot sizes, an automation of the packaging step has been introduced. However, automating the packaging process will decrease the production cycle time (and thus costs) also for mixed variant production lines, thus allowing that several production lines can be merged to a reduced number of flexible packaging stations. This also allows an optimization with regard to the actual demands of the (various) goods (i.e. number of items produced per day).

Within the CustomPacker project [1], the goal is to design and assemble a packaging workstation mostly using standard hardware components resulting in a universal handling system for different products (size, weight and form). Ideally a setup for packaging a high variety of products and components can be implemented. The end-user in the CustomPacker is Loewe AG and will provide different TV-sets for the different packaging tests and use cases.

P. Neto and A.P. Moreira (Eds.): WRSM 2013, CCIS 371, pp. 158–169, 2013.

This paper is organized as follows: section 2 gives an overview of different grasping and packaging approaches used in the industry and research. In section 3 the different technologies that are identified for the packaging process in the CustomPacker project are presented and described how these technologies are developed and applied in the process. Section 4 shows first experimental results of the packaging process and finally section 5 summarizes and concludes the paper including further research approaches.

2   State of the Art

2.1   Packaging

Packaging has been a huge part of the production industry. Many (consumer) products are delivered to the customer in a packaged form. Many different packaging methods are available on the market nowadays, from using paper or foil as packaging material up to large cardboard boxes that are used to protect the products against damage during transportation.

Many packaging approaches are applied in the industry nowadays. In many companies the human workforce is still very large, especially in low wage countries. People are standing along conveyor belts and placing the objects inside their package. As this work is very tedious and in many situations also not very ergonomic, more and more countries are moving towards automated packaging. In 2010, the number of robots for packaging increased considerably from 232 units to 653 units, accounting for a share of 4% of the total supply of robotic units sold [2]. Nowadays, all robotic suppliers are providing their own solutions for robotized packaging. One of the most known packaging systems is the FlexPicker^{TM} developed by ABB [3]. Much research has been performed in optimizing the packaging process, where mostly the focus has been on the grasping of the products to be packed. In [4], the focus on the picking of flexible objects, like food. The system tracks the flexible objects and plans the path of the robot so that it can calculate how the object will move at the end of the conveyor belt and to robot is able to grasp it. Other approaches focus on the use of Virtual Reality for the grasping of different objects [5][6].Problem with many of these solutions is that they are only suitable for one specific product family, or even if different versions of the same product (e.g. different TV-sets) need to be packed, the system needs to be modified.

2.2   Flexible Gripping

To grasp objects with different shapes with the same gripper, its adaptability should be enhanced. In order that the gripper is more adaptive, the two main approaches used are either to increase the active number of degrees of freedom (DOF) of the fingers by adding joints and change the position of the fingers, or create passive degrees of freedom with deformable fingers to adapt to the shape of the object. Most of these devices rely on mechanical gripping technologies. The problems in both cases are to keep the fingers stiff and strong enough to ensure the gripping force, to ensure the reliability of complex mechanical designs and being able to automate the motion planning.

1) Adaptability by means of deformable end-effector
The deformable grippers adapt to the shape of the object, not with mechanical joints but with the deformation of the material. This creates flexible joints as passive degrees of freedom.

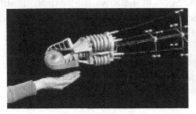

Fig. 1. Deformable end-effector from FESTO and Brown

An adaptive gripper finger is based on the Fin Ray Effect® that has been created by FESTO [7]. The structure of the finger is made of two deformable struts joined at their tips and linked in between by supports. Thus, the finger deforms in the opposite direction of the force application so the rounded object are closed. The payload is about 5kg.

A revolution in the gripping technologies is the development of the so called universal grippers. They consist of an elastic material filled with granulate material or powder, like coffee powder. The balloons are mounted to the robot and fit themselves to the shape of the object to be picked. By means of vacuum the air gets out of the balloon and the powder hardens, with the result that the object is gripped. To loosen the object, air is pumped in the balloon. With such a gripper, flat objects as well as soft objects and object with a complex shape can be handled. In addition, several objects can be gripped at the same time, while maintaining their relative position and orientation [8][9].

2) Adaptability by means of numerous degrees of freedom of the fingers
The other alternative to increase the adaptability of the gripper is to keep stiff links but increase the number of joints. The most extreme developments that follow this approach have derivated in anthropomorphic hands.

The anthropomorphic hand possesses more than two fingers (generally 3 and 6). Each finger has a certain number of active degrees of freedom, based on mechanical revolute joints, the optimum of joints is usually considered to be three for each finger. In this type of gripper, the adaptability is proportional to the number of actuated joints, i.e. to the number of motors. The actuation of more joints makes it possible to grasp objects of very different shapes and size, by repositioning the fingers. For instance, the adaptive gripper developed by Robotiq [11] and shown in Figure 2, has 3 fingers of 3 DOF. One of the fingers is fixed while the two other are on a yaw joint to ensure different grasping modes. The maximum payload is about 10 kg in encompassing grip.

Another way of repositioning the fingers, as developed by Schunk in its 3-finger dexterous hand [10], is to still have one fixed finger while the other two are mounted on centered revolute joints, as shown in Figure 3.

Fig. 2. Robotiq Adaptive Gripper S-Model

Fig. 3. Robotiq Adaptive Gripper S-Model

3   Technologies

Within the CustomPacker project [1], the goal is to develop a flexible packaging system. The system consists out of different components that will optimize the packaging process. In the upcoming sections, three different components of the proposed solution will be presented, which will cover the topics of object pose recognition, for detecting where the object to be grasped is located and in which pose, the path planner for moving the gripper to the object to be grasped and finally, the flexible gripper that is being developed within the project.

3.1   Object Pose Recognition

The object pose recognition relies on a 3D surface scan of the object to be grasped. Therefore, the software ReconstructMe [15] is used to generate an accurate 3D surface model of the object to grasp. The ReconstructMe system consists of a depth image input device, such as the Microsoft Kinect or Asus Xtion, a computer for the calculation of the 3D surface and visual feedback to the user. For 3D surface reconstruction, the user takes the depth image capture device in his hand and films the object from different viewpoints. The 3D surface model is captured in real-time, at approximately 30 fps. The real-time capability is achieved through calculation on the GPU, which provides data parallel execution.

The theoretical background of Reconstructme is to construct a copy of the surface via a truncated signed distance function (TSDF), represented by a union volume grid with a configurable size in both dimension extensions (x, y, z) and grid size. Thus, the 3D reconstruction is limited to the size of the volume, and the accuracy can depend on the grid size. Each depth image frame of the input device is integrated via the TSDF into the volume. Since the camera can move around freely, the position of the camera has to be tracked relative to the volume. Therefore, the previous depth images and the current one are transformed to point clouds in the camera coordinate system and aligned by a high speed version of the Iterative Closest Point (ICP) algorithm. Once the reconstruction is finished, a triangulated model can be exported by the marching cube algorithm [13].

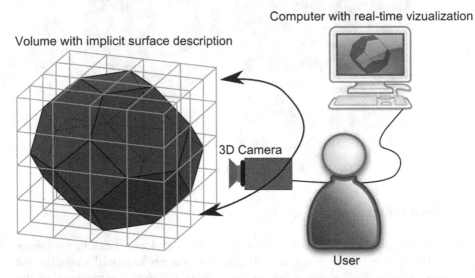

Fig. 4. Setup for constructing CAD models of unknown objects with ReconstructMe

As mentioned, the company Loewe AG is the end-user in the research project and provides various versions of TV sets for testing the packaging process. For gripping TV sets, the real 6 DOF pose of the TV set needs to be estimated. Since TV sets are generally black and featureless and are additionally presented in roughly the same position each time, ReconstructMe itself is used to determine the pose of the TV set in the following way.

First, a model of each TV set is generated using ReconstructMe. The volume containing the reconstruction is saved to the database. Next, when a match is required, the corresponding volume model is loaded from the database into an instance of ReconstructMe. Then, the current sensor input is matched against the content of the volume using a standard ICP [14] algorithm. This procedure is repeated until the sensor view and the content of the volume are similar enough, in which case the pose of the TV set with respect to the volume content is known. TV sets tend to sway when they are moved towards the packaging station. This is the reason why the TV set is tracked in real-time, until it comes to a rest and can be safely gripped.

The current system required a one-time establishment between the content of the volume and the CAD model of the TV set. The intention is to avoid this step in a later stage of the development, by allowing a direct import of the CAD models into the reconstruction volume.

Fig. 5. Reference scan of the TV-set with ReconstructMe

Fig. 6. Result of the object pose recognition

3.2   Flexible Gripping

The adopted concept for the flexible gripper consists on a hybrid solution between deformable end-effectors and grippers with a high number of DOFs.

On the one hand, three independent linear stages provide adaptability to very different size parts, from under 100 mm to up to 1000 mm. Moreover, these stages enable the gripping in four different directions, so that gripping can be performed from both the outer and the inner side of the parts. On the other hand, a very compliant contact area implemented passively with an elastomeric pad assures good adaptation to small features of the part.

Fig. 7. Mechanical arrangement of the flexible gripper

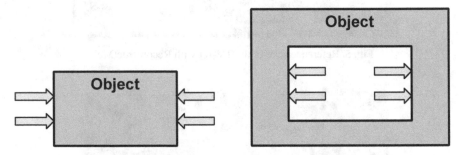

Fig. 8. Gripping from outside and from inside

Apart from the mechanical arrangement and features of the gripper, the controlling possibilities enable the use of the gripper with very different parts. Mechanical characteristics of the object greatly determine the suitable gripping technology. In the following, two control possibilities are being described that are provided by the developed gripper.

In Force Mode, the gripper will apply the commanded force against the parts with the fingers. Figure 9 shows the gripping phase in this mode. When gripping, the fingers move at a high speed until they reach a given position. From that position on, the speed is reduced because contact with the part is expected and the fingers continue

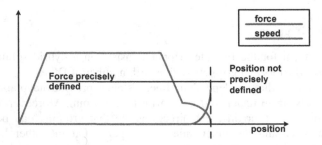

Fig. 9. Gripping in Force Mode

moving until the commanded contact force is fully applied against the part. The precise final position of the fingers is not defined by the user, but the object shape and stiffness will define it. The recommended grasping force depends on the weight of the object. The release of the object is much simpler. The fingers start to move until they reach the opening speed and continue moving until they reach the final position.

In position mode, the fingers move to the commanded position. When gripping, the fingers move at the commanded speed to the commanded position. This working mode is very appropriate to handle very flexible and lightweight parts, such as empty carton boxes. The applied gripping force is not well known, but it is defined by the deformation that the part suffers and its stiffness. Figure 10 shows the gripping phase in position mode.

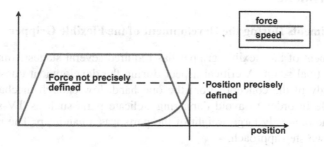

Fig. 10. Gripping in Position Mode

The manipulation planner creates the collision-free path needed for the robot to move its arm to the recognized object and to complete its tasks: gripping and depositing an object with colliding with the static environment. The planner is further responsible for creating the path needed for simulating the robot movements. The calculated path is also translated into the robot-specific programming language.

The planner first collects all the grip poses of each recognized object (each part is prepared offline) and the inverse kinematics for each object is computed using an Inverse Kinematics (IK) solver. The planner iterates on all grip poses until all path checks are collision-free by checking the geometries of the robotic working cell by using simplified CAD models. The following checks are performed for collision-free paths:

- Verification if the IK solution of the grip pose is collision-free
- Determining if the direct path from the home position to a fixed distance above a grip pose, to so called pre-grip point, is collision-free
- Determining if the direct path from the pre-grip point to the grip point is collision-free
- In the last step, the planner checks also if the reverse path (same as the grasping path only vice versa) is collision-free with the difference, that the grasped object is now dynamically attached to the gripper.

When a collision-free path is found, the path planning algorithm writes a robot-specific movement file (based on different templates for many types of robots) which is ready to be executed by the robot. The manipulation planning for this research is based on a prototype of a compliant SCARA robot, developed by Ferrobotics [15] for safe human-robot-collaboration with only four degrees of freedom. Thus the robot's reachability is very constrained because the gripper cannot have a transformation with an x- or y-rotation (with respect to the robot base). The planner disregards these rotations by decomposing the transformation of the flange and the grip point into an Euler zyx-rotation and by discarding the x- and y-rotation. Thus the orientation of the grip-point is in every case.

4   Experiments

4.1   Experiments During the Development of the Flexible Gripper

The development of the flexible gripper has required several stages from conceptual design to the final setup. A critical aspect during the development consisted on the feasibility study of the concept. From the one hand, low enough mechanical forces were advisable in order to avoid damaging delicate parts such as TV-sets. On the other hand, the relatively large weight of the parts was a handicap. The problem was handled in a two-step approach.

Firstly, a comprehensive tribological study was carried out in order to obtain friction coefficients of a bunch of elastomers with different surface finished against a set of counter-materials that are commonly used for the housing of consumer goods. The study required an Optimal Model SRV Tribometer to evaluate friction and wear characteristics in laboratory conditions. No noise, high static and dynamic friction coefficient and low wear were the criteria followed for the selection of the best tribo system. The results revealed that some materials such as EPDM could work appropriately.

Later, a test was designed to try the gripping of parts in the same mechanical configuration as in the final gripper. The test bench trials were successful and showed that 100N to 200N were enough to safely grip TV-sets like parts up to 30kg in weight. The driving forces that have been followed during the design of the flexible gripper have been the functionality, the safety and the lightweight construction. The gripper has been designed to work in the vicinity of people. Several design decisions have been taken in order to reduce risks. From the point of view of passive security, soft and lightweight materials have been used to reduce impact forces. On the other hand, active impact safety schemes have been adopted for the main mobile parts of the gripper. The main structure is monitored by means of a certified "bumper" that will stop both the gripper mobile parts as well as the manipulator where the gripper is mounted. In order to reduce the mass of the gripper, extensive use of carbon fiber and aluminium has been made. Moreover, design optimization by Finite Element Analysis has been critical in order to avoid unnecessary weight.

4.2   Adaptability of the Gripper to Small Features of the Part

The tried finger shape has adapted very well to different TV-set shapes. The 25mm long deformable pad area has been enough to appropriately embrace both curved and flat surfaces. The following Figure 11 shows the fingers close around three different TV-sets.

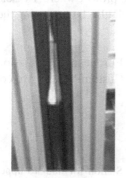

Fig. 11. Compliant pad adaptation to small features of the object

4.3   Gripper Performance Tests

Several tests have been carried out to check the gripper performance. On the one hand, the available strokes and distances between the fingers have been checked, resulting in 140mm and 820mm strokes for short and long stroke stages respectively. Gripping force measurements have been carried out locking force transducers between the fingers. The obtained results for maximum continuous force are in the 540 +/- 20N range. On the other hand, force control within a 3% of the maximum forces has been achieved, which enables fine force control for gripping applications. The reliability of the gripper has been checked by means of ling time gripping tests. Trial of around 5 hours part holding have been successful. Finally, the gripper has performed successfully gripping lightweight parts like a 40x300x200 mm cardboard box, as well as 30kg and 46" TV-sets.

5   Summary

The packaging industry is looking more and more at approaches for optimizing the packaging process and providing humans with ergonomic enhancements. This paper showed results from the CustomPacker project where the goal is to create a highly customizable and flexible packaging station. A cheap but effective object pose recognition solution is introduced where the object to be grasped is digitally constructed and detected using a Microsoft Kinect. With regard to the developed gripper, it is shown that it performs appropriate with very different sized parts, and that it can handle strong and heavy parts as well as delicate and lightweight ones. The adaptability has been provided by a hybrid mechanical solution comprising three linear stages and compliant pads supported by a double mode control strategy consisting on position of force mode control. The CustomPacker project has now reached the integration phase where the different modules are being merged together for the packaging process. First tests have already been performed for the integration. Finally, the system needs to be integrated at the end user of the project where TV-sets will be packaged. During this phase of the project, the focus will also be on testing the solution on different kind of products, like e.g. solar panels.

Acknowledgments. This work has been supported by the European Commission under the seventh Framework Programme within the CustomPacker project (FoF.NMP.2010-1-260065).

References

1. CustomPacker – Highly Customizable and Flexible Packaging Station for mid- to upper sized Consumer Goods using Industrial Robots, http://www.custompacker.eu (access Date March 2013)
2. World Robotics, Industrial Robots. Published by the IFR Statistical Department, hosted by VDMA Robotics + Automation, Germany (2011) ISBN 978-3-8163-0635-1
3. ABB FlexPicker™, http://www.abb.com/product/seitp327/cf1b0a0847a71711c12573f40037d5cf.aspx (access Date March 2013)
4. Jordt, A., Fugl, A.R., Bodenhagen, L., Willatzen, M., Koch, R., Petersen, H.G., Andersen, K.A., Olsen, M.M., Krüger, N.: An Outline for an Intelligent System Performing Peg-in-Hole Actions with Flexible Objects. In: Jeschke, S., Liu, H., Schilberg, D. (eds.) ICIRA 2011, Part II. LNCS, vol. 7102, pp. 430–441. Springer, Heidelberg (2011)
5. Pichler, A., Ankerl, M.: User Centered Framework for Intuitive Robot Programming. In: International Workshop on Robotics and Sensors Environments (ROSE), Phoenix, Arizona, USA, October 15-16 (2010)
6. Pichler, A., Wögerer, C.: Towards Robot Systems for Small Batch Manufacturing. In: IEEE International Symposium on Assembly and Manufacturing (ISAM 2011), Tampere, Finland, May 25-27 (2011)
7. Festo Corporate – BionicTripod with FinGripper, http://www.festo.com/cms/en_corp/9779.htm (access Date March 2013)
8. Anne, Y.: Balloon filled with ground coffee makes ideal robotic gripper, October 25. Cornell University, Chronicle (2010), http://news.uchicago.edu/article/2010/10/25/balloon-filled-ground-coffee-makes-ideal-robotic-gripper-research-shows
9. Amend, J.R., Brown, E., Rodenberg, M., Jaeger, H., Lipson, H.: A Positive Pressure Universal Gripper Based on the Jamming of Granular Material. IEEE Transaction of Robotics 28, 341–350 (2012)
10. SCHUNK – SDH, http://www.schunk-modular-robotics.com/left-navigation/service-robotics/components/actuators/robotics-hands/sdh.html (access Date March 2013)

11. Dexterous Robot Gripper – Adaptive Gripper,
    http://robotiq.com/en/adaptive-gripper/ (access Date March 2013)
12. ReconstructMe, http://www.reconstructme.net (access Date March 2013)
13. Lorensen, W.E., Cline, H.E.: Marching Cubes: A high resolution 3D surface construction algorithm. Computer Graphics 21(4) (July 1987)
14. Besl, P.J., McKay, N.D.: A Method for Registration of 3-D Shapes. IEEE Transaction on Pattern Analysis and Machine Intelligence 14, 239–256 (1992)
15. FerRobotics – Robotis with physical compliance. FerRobotics Compliant Robot Technology GmbH, http://www.ferrobotics.at (access Date March 2013)

Vision-Based Automation of Laser Cutting
of Patterned Fabrics

Anton Garcia-Diaz[1], Isidro Fernández-Iglesias[1], Enrique Piñeiro[1], Ivette Coto[1],
Félix Vidal[1], and Diego Piñeiro[2]

[1] AIMEN Technology Center, R/ Relva 27A,
36410 O Porriño, Spain
{anton.garcia,isidro.roberto,epineiro,mcoto,fvidal}@aimen.es
[2] SELMARK, PTL Valadares R/ C Nave 11,
36315 Vigo, Spain
diego.pineiro@selmark.es

Abstract. A machine vision system was developed to work in combination with
a laser robot cell, in order to fully automate cutting of patterned and deformable
fabrics. The system exploits shape-based matching for online CAD-lace
alignment in robot coordinates. Also, it enables an extremely easy programming
very similar to current manual alignment practice, but done in a graphical
environment and only once at the beginning of the production of each new
reference. The results obtained in lace cutting are compliant with existing
quality standards and thus support the validity of the proposed approach. The
major benefits are the suppression of die cutters thanks to the use of laser
technology and the automation of manual low value repetitive tasks.

Keywords: laser cutting, lace cutting, shape models, matching, machine vision.

1   Introduction

Laser technology is available nowadays for cutting of fabrics, and even to reliably
draw decorations in plain and deformable fabrics. As compared to traditional cutting
methods, laser cutting offers several advantages such as fast cutting speed, reduced
time consumption, a non-contact cutting and no tool wear [1]. Therefore, the use of
this technology improves efficiency and productivity compared to conventional
cutting methods.

Image processing is already used to locate cut marks in certain laser machines[1] and
local approaches using magnifying objectives or laser stripes to track and measure in
real time thread thickness and shift have been demonstrated to control laser heads
(process parameters and trajectory correction) [2].

Otherwise, CAD-based approaches have been demonstrated for different industrial
applications to facilitate robot programming [3][4]. Moreover, the FP7 LEAPFROG

---

[1] http://www.gerbertechnology.com/en-us/solutions/
technicaltextiles/cutting/low-ply/contourvision.aspx

P. Neto and A.P. Moreira (Eds.): WRSM 2013, CCIS 371, pp. 170–178, 2013.
© Springer-Verlag Berlin Heidelberg 2013

project has demonstrated the usefulness of CAD-based robot programming to facilitate the automation of sewing operations [5].

However, the automation (e.g. using existing laser robot cells) of cutting pieces on patterned and deformable fabrics (e.g. laces), which is a common operation in different textile industry segments has not been solved yet. The main issues are related to fabric positioning, that is to alignment in robot coordinates. The high deformability and variability (in size and shape) of this kind of fabrics pose very tough requirements on fixtures. Moreover, the use of robots poses the need of enrolment (or training) of personnel skilled in software for robot programming (with teach pendant or CAD-based offline programming software). Existing programming approaches do not allow to easily defining the cut shape on different lace fabrics, in a similar way that textile operators currently do for positioning die cutters on the laces.

Fig. 1. Current manual die cutter positioning for cutting pieces of lace at Selmark

The main use of machine vision in textile has focused on defect detection and texture identification for quality control, with few attempts of vision guidance for process automation [6]. A recent exception may be found in the FP7 ROBOFOOT project, which aims to apply 3D machine vision to shoe localization for grasping and manipulation [7].

This paper tackles the automation of lace cutting for lingerie manufacturing. This process is currently done through semi-automatic procedures. The **Fig. 1** shows a cutting operation in which a skilled worker manually places a die cutter on the lace, and next drives a cutting machine to extract a lace piece.

The fully automatic approach here proposed combines a very simple CAD-based and graphical offline robot programming with vision guidance for laser cutting. To this end, a shape-based lace localization system is proposed that allows firstly the extraction of a visual model of the decoration aligned to the cut trajectory, and secondly its accurate localization (position and orientation) in subsequent images for successive piece cuts. A prototype of the system has been installed and assessed in a laser robot cell, cutting actual pieces of lace from a lingerie collection. Preliminary results show that the system is robust to illumination conditions and severe lace misalignment (due to both shifts and rotations) and that the resulting pieces comply with aesthetical requirements.

This paper is organized as follows. In section 2 the system is described in detail. Section 3 addresses the assessment of the prototype based on a selection of different laces and pieces from an actual lingerie collection. Finally, section 4 highlights the main conclusions.

2   System Description

2.1   Prototype Setup

The prototype developed used a low cost GigE vision industrial camera of 1360x1024 of resolution. The camera was mounted on ceil position inside a VotanC laser cutting robot cell from Jenoptik AG [2]. The main components of the cell are a IRB-2400/16A ABB robot, a CO_2 Rofin Sinar SCx30 laser with a maximum power of 310W and 10,6μm wavelength, and a LASERMECH cut head with a focus distance of 127mm and a 2mm diameter nozzle. Since the ambient illumination of the cell (vertical fluorescent lamps at the back side) was found to be enough, no additional sources were used. The outline of the setup and a view of the laser robot cell used are shown in the Fig. 2.

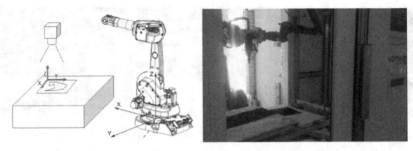

Fig. 2. Left: Scheme of the vision system setup in a laser robot cell. Right: Image of the VotanC laser robot cell used, with a view of one of the two pairs of fluorescent lamps used as illumination.

The camera-robot system was calibrated following a 4-point procedure, under the assumption that every points lay on the same plane. Four points on the working plane and within the camera field of view served as fiducials. Both camera and robot coordinates were acquired at that points and the projective transformation matrix was estimated through the direct linear transform algorithm [8]. Hereafter, the resulting matrix is used to transform image points to robot coordinates. This step needs to be repeated only if the relative position of the camera, the working table, or the robot changes. Since only a reduced field of view was used (aprox. 22°), lens distortion was negligible and no lens calibration was required.

2.2   Programming Approach

Each piece to cut is programmed once for the corresponding lace, following a simple offline procedure prior to its automated production. Firstly, the design technician will have to draw the CAD of each piece, in the same manner that is currently required to order new die cutters. The only additional requirement will be to place the origin of coordinates at a predefined matching point.

---

[2] http://www.jenoptik.com/en-laser-machines-laser-cutting-complex-3d-components-votan-c-bim

Fig. 3. Left: CAD of a piece to cut with the origin defined at the match point. Right: Application for automatic generation of a robot program from the CAD, with an extremely simple interface (DXF file explorer).

From this CAD description of the trajectory to cut, a robot program template is generated. An own developed program takes all the decisions on robot axis configuration and in the mapping from polylines to trajectory points, as shown in the Fig. 3. The exact procedure followed in the conversion from CAD to robot instructions has been described in detail in a previous work [9]. As a result, all the points are referred to the Cartesian coordinates defined in the CAD.

Secondly, the operator of the machine has to manually place the CAD on an image of the lace as shown in the Fig. 4. This approach resembles the current practice of die cutter positioning but on a graphical environment.

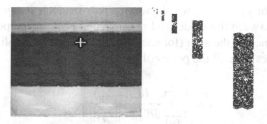

Fig. 4. Left: Manual positioning of the CAD on an image of the lace to cut for model-lace alignment; Right: Example of automatically extracted multiscale shape model of the region of interest (associated to a manually aligned piece) for 0° orientation.

From this manual alignment, the system extracts a region of interest (ROI) around the CAD draw. From this ROI, a multiscale shape model is extracted that will allow the detection and acurate location of the same shape in subsequent images of new segments of the same lace roll. Within the context of this work, a shape-based model approach presents two major advantages:

1. Compared to correlation-based methods, it is robust against illumination changes and presents lower computational loads.
2. Compared to approaches based on keypoints (e.g. SIFT [10]), it does not fail when faced to sparse edges or highly repetitive texture, a requirement to work with decorated laces.

The typical disadvantage of shape-based methods is that they are not invariant to deformations like perspective distortion or material deformation. Since perspective remains constant in the proposed setup, the first is not a problem. Regarding the second issue, it means that where lace deformations are important the model will not be found. As a result, important lace deformations will produce waste of the lace in the affected area.

Therefore, using the ROI that follows the manual alignment of the CAD and an image of the lace, a shape-based template was extracted following the procedure described in [11]. It involved an edge detection so that the model consisted of a collection of n points: $p_i = (x_i, y_i)^T$ and a corresponding direction vector $d_i = (t_i, u_i)^T$ for each point $i = 1, \dots, n$. This procedure was repeated for up to four levels of an image pyramid and for a number of orientations obtained through subsampling and rotation of the extracted model image. To speed-up the process, the orientations were evenly distributed over a limited range ([-50°, 50°]) with limits far beyond of the worst scenario conditions. The specific number of orientations was automatically determined depending on the area of the model.

2.3   Lace Alignment and Robot Guidance

During its online operation, the system acquires an image of the lace and searches the shape model on it. To this end the image is processed again through an edge detector and a direction vector $e_{x,y} = (u_{x,y}, w_{x,y})^T$ is extracted for each image point. Then, a similarity measure between the model and the image at each image point is obtained. Similarity s was defined as the sum (for each image point) of normalized dot product of the direction vectors over all the points of the model, as follows

$$s = \frac{1}{n} \sum_{i=1}^{n} \frac{\langle d'_i, e_{q+p} \rangle}{\|d'_i\| \cdot \|e_{q+p}\|}$$

Where d'_i denotes the direction vectors after an affine transformation of the model accounting for a translation to the q point and a linear transformation $p'_i = A p_i$. This approach ensures robustness to both occlusion and clutter. Moreover, the maximum possible score of a detected object is roughly proportional to its visible area.

The search proceeds through the pyramid from the top level for all possible poses of the model. Local maxima of s over 0.3 were taken as potential matches and were tracked through the subsequent pyramid levels (checking that $s > 0.3$) until they were found at the lowest level. Consequently, the search space is greatly reduced, since only small regions in finest scales (around previous matches in coarse scales) are searched. This gain in search efficiency is the major benefit of the adopted multiresolution approach, rather than scale invariance. Indeed, apparent scale is not expected to vary since depth is kept constant.

Once the model is found in the finest scale, the pose of a recognized model is refined by fitting the similarity to a second order polynomial on the neighborhood of the maximum score. This final step endows the system with subpixel accuracy.

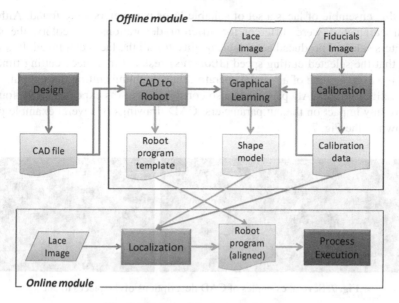

Fig. 5. Scheme of the proposed system

The Fig. 5 shows a complete scheme of the proposed system.

3   Prototype Preliminary Assessment

For the assessment of the prototype, 10 different types of lace decoration and 17 different combinations of decoration and color were used. Six examples of the laces used are given in Fig. 6. Note that even for the same lace, thread width varied between 500-750μm. All the laces were made of nylon.

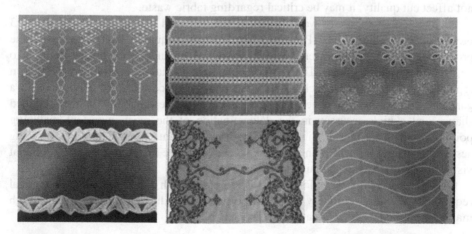

Fig. 6. Six examples of the laces used for assessment of the proposed system prototype. The selection included different sizes, decoration patterns, and colors.

For this ensemble of laces a set of suitable laser parameters were found. Although optimal parameters were different for different decorations and colors, the set of parameters selected produced acceptable quality for all the laces employed. It is worth noting that the selected cutting speed (100mm/s) resulted in a piece cutting time that halved the average time of a manual operator, thus doubling cutting throughput.

As well, different CAD parts were also considered but (as expected) were found to not have any impact on the cut parameters. CAD drawings of several example pieces are shown in the Fig. 7.

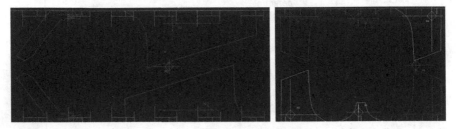

Fig. 7. Several examples of CAD descriptions of the pieces to cut

A preliminary assessment was done twofold. Firstly, laces were moved on the image plane (translations and rotations) and the piece to cut was searched again and again. In all the cases that the pattern corresponding to the piece was found, the error in the position of the found matching point respect to a manually determined position was better than 0.5mm. The Fig. 8 shows an example of matching point localization. The number of false negatives (potential cuts not detected) was very low. However, in this preliminary assessment we cannot provide a definite value of actual false negatives since the lace was manually positioned on the table. This is very important since observed false negatives were observed when wrinkles of the lace were moderate. Therefore, although the position choice in a final industrial setup should not affect cut quality, it may be critical regarding fabric waste.

Secondly, a series of cuts was done using the proposed system. The Fig. 8 shows 3 pieces done in these trials. Personnel specialized in quality control assessed the produced cuts. The criteria were basically visual acceptance of the pieces in order to be used in garment making. This visual acceptance is mainly determined by pattern repeatability and a good matching between pairs of pieces that are contiguous in a garment. The results obtained showed an excellent repeatability in the obtained shapes and a high accuracy in the determination of the most important matching points. This is a key result, since the visual aspect of matching points has a strong aesthetical effect. In these points, different pieces of lace match in a garment, so that visual continuity should be maintained.

Overall, the results were found fully compliant with the demanding aesthetical requirements, those currently imposed on actual industrial production through manually aligned (semi automated) cut processes.

Fig. 8. Left: Example of a detected shape model and detail showing the estimated position of the matching point with subpixel accuracy. Right: Cutting process during a trial and example of a series of 3 automated cutting operations of the same piece.

4   Conclusion

In this paper, a machine vision system that enables the automation of cutting operations on patterned fabrics has been demonstrated. The system allows a fast graphical programming of cut trajectories in a PC through drag and drop operations of the CAD of the pieces on an image of the fabric. These operations resemble the manual positioning of die cutters, currently used in industrial production. Relying on this graphical alignment, the vision system locates the next useful region of fabric for cutting and realigns the CAD of the cut in robot coordinates. From this information, the robot program (and trajectory) is automatically generated.

As a result, a single and simple graphical alignment operation replaces the manual positioning currently used in factories for each manufactured piece. Therefore, each piece is programmed once and from there on, the full production may be automated using a laser robot cell, free of manual operations.

The major potential industrial benefits of the system are the following

- Minimum of 50% of reduction of peak cutting time
- Increase in production capacity thanks to full-time cutting (no interruption; 24h production), without higher labor costs
- Bearing in mind the estimated operation costs of a laser robot cell (half of the current costs of a manual operator), and the increased throughput, the laser cutting productivity may increase by a factor of 4
- Strong reduction of consumables, thanks to the suppression of die cutters, between 300-400 units per year depending on collections for a manufacturer like SELMARK
- Enabling manufacturing of single trial designs at virtually no additional cost

Although a full series of trials using an actual industrial lace feeding system with more different piece models and laces is still required for pre-industrial validation, the system has succeeded in cutting laces for lingerie garments. The fabrics and pieces used pose important challenges due to a high deformability and tough aesthetical requirements. Such requirements imply in practice that submilimeter accuracy is needed for certain matching points.

The shape-based approach adopted has shown robust to contrast conditions resulting from both illumination and lace color. Indeed, no specific illumination other than pre-installed robot cell lamps has been used and lens aperture was kept constant for laces of different colors (ranging from black to white). However, a large-scale assessment to characterize the reliability of the system with a full collection of different fabrics is required to confirm that no specific illumination is required in actual production conditions.

Future work will address the adaptation of trajectories to local deformations of fabrics (using deformable models) in order to minimize fabric waste, and the segmentation of the lace in order to use specific laser parameters depending on local thread width, to ensure quality in spite of strong thread width variations.

Acknowledgments. This work has received financial support from the Xunta de Galicia and FEDER through the SALMON project.

References

1. Yusoff, N., Osman, N.A.A., Othman, K.S., Zin, H.M.: A Study on Laser Cutting of Textiles. In: Proc. of International Congress on Applications of Lasers & Electro Optics (2010)
2. Bamforth, P.E., Jackson, M.R., Williams, K.: Transmissive dark-field illumination method for high-accuracy automatic lace scalloping. International Journal of Advanced Manufacturing Technology 32, 59–60 (2007)
3. Neto, P., Mendes, N., Araújo, R., Pires, J.N., Moreira, A.P.: High-level robot programming based on CAD: dealing with unpredictable environments. Industrial Robot, Emerald 39(3), 294–303 (2012)
4. Neto, P., Mendes, N.: Direct off-line robot programming via a common CAD package. Robotics and Autonomous Systems (in press 2013)
5. Walter, L., Kartsounis, G.-A., Carosio, S.: Transforming Clothing Production Into a Demand-driven, Knowledge-based, High-tech Industry: The Leapfrog Paradigm. Springer (2009)
6. Ngan, H.Y.T., Pang, G.K.H., Yung, N.H.C.: Automated fabric defect detection - A review. Image and Vision Computing 29, 442–458 (2011)
7. Maurtua, I., Ibarguren, A., Tellaeche, A.: Robotics for the benefit of footwear industry. In: Su, C.-Y., Rakheja, S., Liu, H. (eds.) ICIRA 2012, Part II. LNCS, vol. 7507, pp. 235–244. Springer, Heidelberg (2012)
8. Hartley, R., Zisserman, A.: Multiple view geometry in computer vision. Cambridge Univ. Press (2000)
9. Álvarez, M., Vidal, F., Iglesias, I., González, R., Alonso, C., Remuinan, M.: Development of a flexible and adaptive robotic cell for marine nozzles processing. In: 17th IEEE Conference on Emerging Technologies & Factory Automation (2012)
10. Lowe, D.G.: Distinctive image features from scale-invariant keypoints. International Journal of Computer Vision 60, 91–110 (2004)
11. Steger, C.: Similarity measures for occlusion, clutter, and illumination invariant object recognition. Pattern Recognition, 148–154 (2001)

Industrial Robots Accuracy Optimization in the Area of Structuring and Metallization of Three Dimensional Molded Interconnect Devices

Arnd Buschhaus and Jörg Franke

Institute for Factory Automation and Production Systems
Friedrich-Alexander-University Erlangen-Nuremberg
Erlangen, Bavaria, 91058, Germany
{Arnd.Buschhaus,Joerg.Franke}@faps.uni-erlangen.de

Abstract. Based on their beneficial features, there is a growing importance of 3D molded interconnect devices. An important process in course of their manufacturing is the conducting pattern generation. Therefore industrial robots can be used, moving the substrate with a high accuracy relative to a process nozzle. Due to complex conducting patterns an offline programming of the robots trajectory is needed. Given to the robot, this leads to a precision of the movement not meeting the product demands. To overcome this problem, the usage of modern sensor systems for process monitoring and real-time robot control is investigated. By supervising the conducting pattern generation the actual robots position is derived. Subsequently correction values are determined and given to the robot control position deviations are compensated. For this way of proceeding an efficient methodology for data processing and correction value determination is developed. First experiments already show good results with high geometric accuracies.

Keywords: robotics, sensor systems, 3D-MID, closed-loop-control.

1 Introduction

Three dimensional molded interconnect devices (3D-MID) are 3D shaped thermoplastic substrates, carrying interconnecting conducting patterns and electronic components. Due to the combination of mechanical and electronic functions, they bear a vast number of benefits compared to conventional electronic assemblies, like a high integration of different functions, a reduced number of parts and a great design flexibility. These advantages lead to a rising usage of 3D-MID for example in the automotive, medical and communication branch in the last years and decades. Based on the possibility of optimally filling a limited installation space, an increasing number of miniaturized MID-products with growing complexity is developed.

P. Neto and A.P. Moreira (Eds.): WRSM 2013, CCIS 371, pp. 179–190, 2013.

Several examples for actual 3D-MID products are shown in Fig. 1.

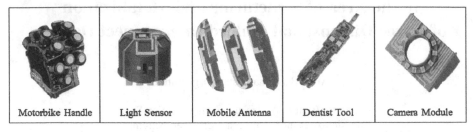

| Motorbike Handle | Light Sensor | Mobile Antenna | Dentist Tool | Camera Module |

Fig. 1. Exemplary 3D-MID Applications [1], [2], [14], [15], [16]

For meeting actual challenges which accompany the miniaturization and increasing integration of functions, sophisticated process technologies have to be provided, making an adequate 3D-MID production possible.

The added value chain of 3D-MID manufacturing can be basically clustered in the following steps: injection molding of the substrate, creation of the conducting pattern layout, application of the solder paste, subsequent placement of electronic components on the substrate and a final soldering process. One essential step forms the generation of the conductive patterns at the geometrically correct places on the different 3D-MID surfaces.

2   Conductive Pattern Generation

For creating the conducting patterns, several technologies are available, which conventionally consist of a structuring process and an additional metallization of the substrate. While during the structuring processes the geometrical routes of the conducting patterns are defined, an application of the plating material (e.g. copper) takes place in course of the metallization. This metallization is realized for example in a chemical galvanic bath. An overview and detailed description of the different state of the art technologies is given in [3].

In summary, it can be stated, that the available standard technologies are well investigated, approved and good process results can be obtained. A general drawback of the conventional technologies is the necessity of using cost-intensive product-specific tools like stamps or masks. Thus, a flexible and fast reaction on different product variants, small batch sizes and prototypes is barely possible and economically not viable.

2.1   Printing Technologies for Flexible 3D-MID Metallization

To overcome this drawback, innovative metallization technologies are developed, which render the necessity of product specific tools for the conducting pattern creation unnecessary. With innovative printing technologies it is even possible to shorten the added value chain, since the conducting pattern material can be directly applied at the geometrically correct places without having a separate structuring process. This is accompanied by a cost reduction of the 3D-MID production, since the need for expensive process equipment can be reduced.

Currently, three main printing technologies for additive conducting pattern generation on 3D-MID surfaces are available: Plasmadust[1], Flamecon[2] and Aerosol Jet Printing[3] . The general function principles of these printing technologies are depicted in Fig 2.

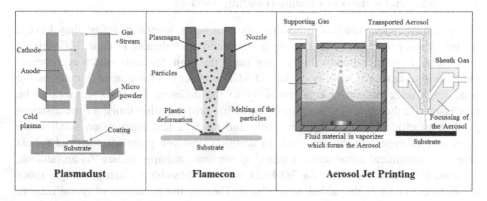

Fig. 2. Printing Technology Overview (pictures based on [4], [5], [6], [12])

These printing technologies have in common, that the plating material consisting of nano and micro-particles is transported by a gas stream and subsequently sprayed onto the substrate. Apart from conventional metallic materials like copper, alumina, silver or gold which are applied by these processes, as well polymers and ceramics can be used [4] [6] [7]. This broad variety of different coating materials enables the printing technologies not only for the generation of conducting patterns, but furthermore the creation of simple electronic components like resistors, capacitors and inductors.

One main challenge, which accompanies these process technologies, is a geometrically exact definition of the process places on the 3D-MID substrates surface, since the coating material – depending on the utilized nozzle – is applied at a very small process spot size of around 100 microns or even below. Thus, very fine pitched structures can be created [4] [8] [11]. For defining the process places, either way the process nozzle has to be guided three dimensionally relative to a static work piece or the work piece has to be moved relative to a fixed nozzle. Actually, to avoid irreproducible influences on the process quality, it is recommended keeping the process equipment static and provide a defined movement of the substrate relative to the process nozzle. This movement has to be performed with a high geometric accuracy for meeting the demands of miniaturized 3D-MID and accordingly

---

[1] Plasmadust trademarked by Reinhausen Plasma GmbH.
[2] Flamecon trademarked by Leoni AG.
[3] Aerosol Jet Printing trademarked by Optomec.

miniaturized conducting patterns and layouts. To enable such a movement, sophisticated handling devices have to be used with a high geometric accuracy and as well a high flexibility of movement. Therefore for example 5-axis CNC-systems are used [9].

An alternative approach for manipulating the substrate relative to the process nozzle is the usage of standard 6-axis industrial robots.

2.2   Industrial Robots as Flexible Handling Devices

Using conventional industrial robots in this area bears several benefits: they have a relatively low price, can be easily programmed and are unbeaten in their flexibility of movement. Due to their movability, they meet the high demands which accompany the extensive design flexibility of the 3D-MID approach. One central demand is the need for a three dimensional accessibility to the different 3D-MID surfaces to be processed. Furthermore, by using modern offline programming tools, a time efficient product specific programming is possible and hence a fast reaction on small batch sizes or prototypes can be achieved. In addition, the low price of industrial robots gives an economical advantage compared to conventional approaches. As an outlook, even multifunctional systems for 3D-MID might be possible by using a single robot for different tasks of the added value chain. Thereby the processes of metallization, solder paste application and electronic component assembly can be united in one machine.

The main drawback of using conventional robots as flexible handling devices in printing systems for 3D-MID is their relatively low "absolute accuracy", which exceeds the high repeatability (typically around ± 10 microns for small robots, e.g. [10]) by a factor of about 10 to 15. The absolute accuracy is especially of relevance for offline generated robot trajectories created with software assistance. In such tools for offline programming a compensation of the individual geometric deviations of the robot, based for example on tolerances in the kinematic chain, is not possible.

To overcome this problem conventionally a manual robot programming takes place considering the robots real geometric deviations. Based on the miniaturized three dimensional conducting patterns of 3D-MID and as a result the necessary complex robot movements, a manual programming is not feasible for this task. Thus software tools for robot programming have to be utilized.

Since lateral deviations and geometric tolerances of the conducting patterns to be applied on the 3D-MID surfaces beyond 100 microns are not acceptable, a suitable methodology has to be developed to increase the accuracy of the robots trajectory.

3   Robots Accuracy Improvement by Utilizing Modern Optical Sensor Systems

As stated beforehand an improvement of the robots accuracy in course of its movement – or at least of the programmed control points - can be reached by a manual reprogramming of the trajectory. Thereby a compensation of geometric deviation is achieved. With this approach a transition from the absolute accuracy to the higher repeatability can be obtained, since the real deviations in the kinematic chain of the robot are considered. Once reprogrammed, the corrected control points of the robots trajectory are approached in the following process cycles with a high accuracy. The high robot repeatability, which can be reached by a manual correction, would be as well sufficient for the 3D-MID metallization process.

This way of proceeding is reasonable for applications having moderate accuracy requirements and a limited number of control points to be corrected. Due to the extensive number of necessary control points in the robots control program for 3D-MID metallization in combination with the high accuracy demands, a manual correction is not feasible for this process.

Therefore, in actual research activities at the Institute for Factory Automation and Production Systems (FAPS), a new approach is researched, utilizing modern optical sensor systems for online process monitoring and an automatic correction of the robots trajectory derived from the sensor data.

3.1   System Architecture

A general structural overview of the investigated system architecture for a closed-loop-control of the robot in course of the metallization process is given in the following Fig. 3.

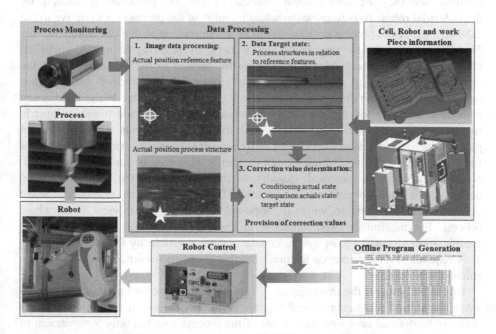

Fig. 3. Online Process Monitoring and Robot Closed-loop-control

An efficient robot programming for complex tasks can be done by utilizing modern robot kinematic simulation tools with the option of generating robot control programs. Therefore, a digital representation of the process cell, the robot and the work-piece is needed as well as additional information regarding process-dependent requirements for obtaining good results. In case of the metallization, process specific and robot relevant requirements are – besides the demanded high accuracy - a preferably perpendicular movement of the substrates surface relative to the process nozzle as well as a continuous speed of this movement. Furthermore, robot dependent constraints have to be taken into account like limited axis accelerations and - to avoid a jittering movement - a restricted number of control points describing the trajectory. Considering all this information, a control program in a robot manufacturer specific source code can be derived. Transferred to the robot control unit an attached robot can move the 3D-MID substrate on the programmed trajectory relative to the static process nozzle, spraying the conducting pattern material onto the surface. This movement is executed with the discussed absolute accuracy, being insufficient for the metallization of miniaturized 3D-MID.

To achieve an improved robot accuracy in actual research activities a high-speed camera system is utilized, monitoring the metallization process in course of its progress to extract the exact position where the coating material is actually sprayed onto the substrate (called "process spot" in course of the next sections). The camera is fixedly mounted with a defined viewing angle next to the process nozzle and records continuously the three-dimensional process, while the substrate is moved by the industrial robot. To determine the actual position of the process spot relative to the 3D-MID, the actual process position has to be set in relation to distinct features of the 3D-MID. In the specific application this is done by referring to marks, so called reference features. These reference features are applied to the substrate before the process starts by using the metallization equipment. A creation of the reference features during the molding process already is not reasonable, due to the insufficient geometric accuracy of the molded substrates. Furthermore additional geometric features of the injection mold would lead to higher tool costs. Hence an exact positioning of these features is done by utilizing a 3D-scanning device. This is part of further research activities, which are not explicitly addressed in course of this paper. Nevertheless it can be assumed, that due to the properties of the surface scanning device to be used an accuracy of around 10 microns for the reference feature generation shall be feasible, which is supposed to be negligible in view of the whole process. For the further explanations it is hence assumed, that during the whole metallization process at least one reference feature is visible by the camera system. This requires several reference features distributed on the substrates surfaces with a suitable methodology.

Due to the setup of the investigated system having a static tool and a static camera while the substrate is being moved relative to these systems, in the cameras image data - in an optimal case - no movement of the process spot but only a movement of the reference features is determinable. By continuously identifying the process spot position and setting it in relation to the reference features of the 3D-MID, the actual robot position is derived. This data is subsequently compared to information of the theoretically correct target state for every time-step, which can be extracted for example from the optimal digital representation of the process based on the kinematic simulation in which as well the product data is considered. As a result, geometric correction values for the robot are generated. This data is subsequently transferred in real time to the robot control and hence a closed loop control is established to compensate position deviations.

The next section addresses the data processing methodology, especially focusing on a time efficient image data processing and the actual state value calculation.

3.2   Data Processing

An overview of the interacting data processing functionalities and modules abstracted from Fig. 3 is depicted in the following Fig. 4.

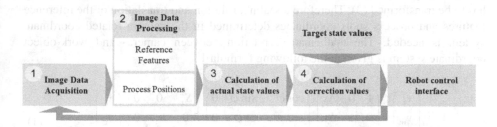

Fig. 4. Main data processing functionalities [13]

As discussed beforehand, due to the system setup, the image data processing is concentrating on the determination of the reference features movement relative to the coordinate system of the camera. In course of the investigations cross shaped reference features have been used, which are detected and described during the image data processing (2). The processing is done by utilizing a combined matching, image data reduction and threshold value approach. In a first step, based on the known shape of the reference feature, a robust shape-based matching and feature detection is performed with a moderate geometric accuracy. For increasing the accuracy, a determination of the reference features center point takes place in a second step. Therefore a region of interest is created around the reference feature detected by the matching process and subsequently a threshold operation is applied in the specified region for extracting the reference features shape. In a third and last step, by a thinning of the detected cross-shaped feature and determination of its intersecting lines, the center point is identified, which can be transformed in numerical coordinates, describing the reference features position relative to the camera coordinate system. Once the initial reference feature leaves due to the movement of the substrate the cameras field of view, another reference feature has to be found and set in a geometric relation to the initial one. Therefore at least in one frame the old as well as the new reference feature have to be visible. During the transition both reference features are detected by the described image data processing operations, their coordinates are extracted and finally the coordinates of the new reference feature are set in relation to the last one. In experiments it has been proven, that this way of proceeding leads to a robust and accurate reference feature description.

Parallel to the reference feature detection, the process spots position has to be extracted from the image data. A time-efficient determination is possible, based on the assumption, that – due to the experimental setup - there is only a very limited movement of the process spot relative to the camera coordinate system. By exploiting this knowledge, a region of interest can be defined in which the process spots position in the acquired image data is expected. In a subsequent step, by a threshold operation applied in this reduced image data, the exact position of the process spot is determined. Furthermore the process spots center point is identified and transformed in coordinates, which can be set in relation to the reference features coordinates to describe the actual state of the process in the camera coordinate system.

For getting the actual state data in the relevant robot or work-object related coordinate system, the information generated in course of the image data processing has to be transformed (3). Therefore a scaling, rotation and translation of the reference features and process spots coordinates determined in the camera related coordinate system is needed. The mathematical relation between camera- and work-object coordinate system is given in the following formula 1.

$$
\begin{pmatrix} x_w \\ y_w \\ 1 \end{pmatrix} = \begin{pmatrix} 1 & 0 & \Delta x \\ 0 & 1 & \Delta y \\ 0 & 0 & 1 \end{pmatrix} \cdot \begin{pmatrix} \cos\varphi & -\sin\varphi & 0 \\ \sin\varphi & \cos\varphi & 0 \\ 0 & 0 & 1 \end{pmatrix} \cdot \begin{pmatrix} S_x & 0 & 0 \\ 0 & S_y & 0 \\ 0 & 0 & 1 \end{pmatrix} \cdot \begin{pmatrix} x_c \\ y_c \\ 1 \end{pmatrix} \tag{1}
$$

In this formula are x_c and y_c = coordinates in the camera coordinate system; x_w and y_w = coordinates in the work-object coordinate system; S_x and S_y = scaling factors; φ = rotation; Δx and Δy translations of the coordinate systems.

The calculation comprises several matrix multiplication operations, whereby the coordinates of the process spot relative to the initial reference feature in the camera coordinate system (x_c , y_c) are transformed in according coordinates relative to the work-object coordinate system (x_w, y_w). The scaling factors S_x and S_y consider the deviation of measures in the image data compared to the real work-object as well as the camera angel relative to the process equipment and are applied for transferring lengths in the camera coordinate system to lengths in the work-object coordinate system. By using the factor φ, a possible rotation of the image-plane around the initial reference feature is taken into account. Finally a translational shift by Δx and Δy is applied to the coordinates. The resulting work-object related coordinates can be directly used as actual state values for the succeeding correction value determination.

For calculating the correction values (4), in addition to the calculated actual state values the target state values have to be taken into account. These have to be extracted from a source with the geometrical and theoretical correct information. Therefore either way the information of the robot control program, the axis-values of the robot or the metallization layout of the product data can be used. In first investigations the correction values are obtained by comparing the actual process spots position for every time-step with the metallization layout of the 3D-MID, which has been transformed beforehand in the same coordinate system as the actual state values with the mathematical operations described already. By this comparison, a distance for every actual and target value is obtained, which can subsequently be decomposed in Cartesian deviation values. These values are used for the closed-loop control of the robot.

4   Experimental Results

Several experiments have been carried out with varying image data processing procedures as well as different camera lens systems (16 mm and 35 mm).A brief summary of the actual results is presented subsequently.

In these first experiments no specialized lighting system has been used, but it has been taken care, that a homogenous illumination of the process space was maintained. By taking a statistically significant number of data processing passes into account, a solid analysis of the geometrical accuracy of the features recognition and description as well as the data processing performance reached so far has been done. Regarding the geometrical accuracy 100 samples have been investigated and for quantifying the necessary time consume of the image data processing 500 samples have been considered.

A good compromise of a high geometrical accuracy of the determined actual state in combination with a relatively large field of view has been achieved with a 5 Megapixel monochrome camera and a 35 mm lens system. With this setup a field of view in the image and hence the processing plane of 41 mm x 55 mm can be obtained, which is sufficient for detecting at least two reference features on the investigated substrates surface during the transition phase discussed in chapter 3.2.

The following Figure 5 shows the 3D-MID being used in course of the investigations with different substrate materials and in different states of the added value chain. For especially investigating the detection and description of the reference features on different faces the part depicted on the very left side of the figure has been used.

Fig. 5. Investigated 3D-MID product

It has to be noted, that the main lateral measures of typical 3D-MID substrates vary in a range of 2 cm (e.g. adaptive cruise control [1]) up to approximately 30 cm (e.g. steering wheel [17]). Accordingly, there is a broad variety of conducting pattern layouts and dimensions of their features available, which can result in the necessity of using varying sizes and distances of the reference features. As conclusion, for some applications other camera and lens system setups might lead to better results.

But, by using the explained image data processing operations and the said 35 mm lens system with the investigated 3D-MID, a detection of the reference feature has been achieved with a mean geometric deviation of 2 microns and below. For the given setup the maximum deviation was found to be 11.4 microns and a standard deviation of 3.3 microns related to the reference features center point has been determined.

For the detection of the process spot in course of the experiments a 16 mm lens system has been used. With this lens system a mean deviation of approximately 40 microns, a maximum deviation of 190 microns and a standard deviation of 60 microns has been obtained. By using as well the 35 mm lens system for detecting the process spot an accuracy improvement by approximately factor 5 can be assumed (Corresponding size to pixel ratio for a 16 mm lens system compared to the 35 mm lens system = 5:1), but even considering this ratio, the reference features are detected with a higher geometric accuracy. An explanation for the difference between reference feature and process spot detection might be the lower color contrast-ratio of the process spot relative to the substrate compared with the high contrast-ratio of the reference features in relation to the substrate. Since in forthcoming experiments the same material for the conductive patterns and reference features is going to be applied and hence as well a high contrast-ratio of the process spot related to the substrates texture is provided, a higher accuracy for the process spot determination can be expected.

Another focus of the investigations has been a minimized time consume of the data processing operations to allow a succeeding real-time control of the robot. With a single reference feature on the substrates surface a steady data processing time per frame of 20 milliseconds and below has been reached. With more complex 3D-surfaces and transitions between different reference features the data processing time is approximately doubled. This is based on the higher complexity of the program: the images with the reference features coordinates have to be saved, the offset has to be calculated, it has to be checked if the reference feature leaves the field of view and if this is the case, a new reference feature has to be detected.

5   Conclusion and Outlook

In this paper a new approach has been introduced, addressing the usage of conventional industrial robots as flexible handling devices in the area of conducting pattern generation of three dimensional Molded Interconnect Devices. For meeting the challenges which accompany the increasing miniaturization and integration of functions in new MID products, a suitable methodology is needed for increasing the robots accuracy. A possible solution proposed in this paper is the usage of a high-speed camera system for inline process control. The information gathered by the sensor system is subsequently used for generating correction values which can be integrated in the robot control to establish a real time closed-loop control of the robots movement. This demanding task requires a high accuracy of the geometric features extraction from the image data in an area of a few 10 microns in combination with an efficient data processing.

In experimental investigations promising results have been achieved already, but further effort has to be put in an enhanced image data processing regarding the accuracy and especially the processing efficiency. One aspect for reaching further improvements is the usage of alternative equipment like a color camera system and a sophisticated lighting device. Thereby a robust feature determination with simplified image data processing operations could be achieved. This might have beneficial effects on the achievable geometrical accuracy as well as the data processing efficiency. Additionally, information which can be gathered beforehand from the MID product data can be exploited further for generating intelligent regions of interest during the image data processing based on the knowledge about expected reference feature and process spot positions in the image data. This will lead to an image data reduction and as a result to a faster data processing performance. For the actual experiments a camera system having a frame rate of 100 fps is used, but in further experiments the usage of a more sophisticated system up to 500 fps is planned. Last an optimized distribution of the processing tasks to different cpu-cores, a framegrabber-card or the graphic board with optimized instruction sets will lead to an accelerated image data processing and hence the properties of the more sophisticated camera system can be exploited.

References

1. Franke, J., Goth, C., Gausemeier, J., et al.: MID-Applikationen. In: MID Studie 2011: Markt- und Technologieanalyse, Erlangen, Germany, pp. 58–68 (2011)
2. N.N.: Effiziente Antennenproduktion mit LDS-Technologies, Garbsen, Germany (2009), http://www.lpkf.de/presse/pressemitteilungen/560/130.html
3. N.N.: Technologiedaten der MID-Herstellungsverfahren. 3D-MID Technologie Räumliche elektronische Baugruppen, pp 149 – 219, Erlangen, Germany (2004)
4. Hedges, M.: 3D Aerosol Jet Printing – An Emerging MID Manufacturing Process. 9. International Congress MID, Nuremberg, Germany (2010)
5. Franke, J.: MIDFlex – MID und Folie. Lecture script MIDFLEX, pp. 46, Erlangen, Germany (2012)
6. N.N.: plasmadust®Technology the revolution in coating technology. Maschinenfabrik Reinhausen GmbH, Regensburg, Germany, http://www.reinhausen.com/desktopdefault.aspx/tabid-1327/1471_read-3696/
7. N.N.: Plasmadust: Elektrische Verbindung und Kontaktierung neu definiert – Verfahren zur chemiefreien Metallisierung und Beschichtung, http://www.epp-online.de/archiv/-/article/32536724/32624659/Verfahren-zur-chemiefreien-Metallisierung-und-Beschichtung/art_co_INSTANCE_0000/
8. Buschhaus, A., Franke, J.: Planning and Control of Robot-Assisted Process Cells for Structuring and Metallization of 3D-MID. 10. International Congress MID, Fuerth, Germany (2012)

9. Reichenberger, M., Jillek, W., et al.: Functionalization of Thermoplastics using Inkjet- and Aerosoljet-Printing Technologies. 10. International Congress MID, Fuerth, Germany (2012)

10. N.N. IRB120 Industrial Robot Datasheet, http://www05.abb.com/global/scot/scot241.nsf/veritydisplay/3bd625bab3c7cae1c1257a0800495fac/$file/ROB0149EN_D_LR.pdf

11. Theophile, E.: Reinhausen Plasma: The plasmadust process: An innovative process for metal coatings on a wide variety of substrates. 9. International Congress MID, Fuerth, Germany (2010)

12. N.N.: Aerosol Jet is not Inkjet. OPTOMEC ADDITIVE MANUFACTURING SYSTEMS, http://www.optomec.com/Additive-Manufacturing-Technology/Inkjet

13. Buschhaus, A., Durst, A.: Kompensation der Ungenauigkeiten von Industrierobotern bei der Strukturierung dreidimensionaler Schaltungsträger durch kameragestützte Bahnkorrektur. unpublished, Erlangen, Germany (2012)

14. N.N.: Technische Informationen: Elektronik – Klimasensoren. Hella KGaA Hueck & Co., http://www.hella.com

15. Knabe, J.: Kombinierter Sonnen-/ Umgebungslicht-Sensor in 2K-MID-Technologie. In: Workshop Innovative Anwendungen der MID-Technik, Stuttgart, Germany (2007)

16. Franke, J., Goth, C., Fischer, C., Pfeffer, M.: Effiziente rechnergestützte Produktentwicklung für räumliche elektronische Baugruppen (3D-MID). In: Zeitschrift für wirtschaftlichen Fabrikbetrieb, pp. 925 – 930, Germany (2009)

17. N.N.: Automotive, http://www.selectconnecttech.com/applications/automotivedesh

Influence of Vibration Induced Disturbances
in an Automatic Inspection Cell

Anna Runnemalm, Tongwei Liu, Mikael Ericsson, and Anders Appelgren

Department of Engineering Science, University West, SE – 461 86 Trollhättan, Sweden
{Anna.Runnemalm,Tongwei.Liu,Mikael.Ericsson,
Anders.Appelgren}@hv.se

Abstract. In the modern manufacturing industry, quality assurance is important. Over the last few years, the interest in automatic inspection has increased and automatic non-destructive testing (NDT) has been introduced. A general automated inspection cell consists of a mechanized system for scanning and a computer system for automatic analysis of the data. In the manufacturing industry, it is preferable to use industrial robots as the scanning equipment since they offer great flexibility, excellent support organization and the in-house know-how is normally high. Another benefit is that a robot can carry different inspection equipment and an inspection cell can therefore include more than one NDT method. For an automatic analysis, high quality of the resulting data is essential. However, a non-stable condition of the NDT sensor mounted on the robotic arm may influence the results. This paper focuses on the influence of the vibration induced disturbances on the results from an NDT system. Vibration amplitude of a point to point robot movement on the robotic arm is measured. The influence of vibration disturbances on the inspection results are evaluated on the thermal images from a thermography system mounted on a six axis industrial robot. The thermal images taken by the system during the movement and after the stop of the robot are evaluated, and the influence of the vibration in these two situations is considered.

Keywords: Automatic non-destructive testing, NDT, Vibration, Thermography, spot weld.

1 Introduction

In the modern manufacturing industry, welding applications (both spot welding and continuous welding) are widely used. For example, more than 4000 spot welds can be found in a typical car body. With an increased requirement for high productivity and competitiveness, achieving full automation is essential for the manufacturing process. Meanwhile, it is of importance to assure weld quality in parallel with high productivity. The interest in automatic inspection has therefore increased.

Non-destructive testing (NDT) can be described as inspection of a product in such a way that the testing does not affect the object's future usefulness. Several NDT methods exist and are usually chosen depending on the properties of the object to be

P. Neto and A.P. Moreira (Eds.): WRSM 2013, CCIS 371, pp. 191–202, 2013.

inspected, which types of different defects should be detected and limitations due to the surrounding environment. NDT of welds is commonly done manually. Automated inspection presents several difficulties, but also advantages such as good repeatability, high speed and stability over a long time period. It will also add an additional criterion for choosing which NDT method to use since not all methods have the same feasibility for automation. Carvalho et al. [1] showed, using three types of artificial weld defects in a pipeline, that automated ultrasound inspection gave a superior result for detecting defects when compared to manual inspection. Even if the advantages of automatic NDT are proven, several questions need to be answered before an automatic inspection cell can be implemented in the manufacturing industry.

In automatic inspection, industrial robots can be used as the scanning equipment [2, 3]. The robots should be able to work with high performance, which means running at high speeds with high accuracy. Nonetheless, for high speed movement, the robot will induce vibration in the device mounted on the robotic arm. Applications carried out by the robots can be affected by the vibrations. Therefore, vibration is an important factor to improve the robot's performance. It has been shown that the variation in load does not have much influence, but the vibration could play a vital role in the performance of the robot [4]. Some works have been done to analyse the vibration, characteristics as well as control, and reduce the vibration. A simulation method for vibration analysis based on virtual prototyping technology is proposed to improve the dynamic performance of a spot welding robot [5]. The practical method is investigated to suppress residual vibrations of industrial robots with an input shaping technique [6]. Based on the vibration suppression control, a robust position servo system for industrial robotics is proposed by Ohishi [7]. In a previous work by Runnemalm [2], the influence of vibration during continuous movement by a single robot was presented. The influence was only briefly studied, and no comparison with differing robots speeds was presented. This paper will complement the previous study.

There exist several non-destructive testing methods [8-10] although not all are suitable for automation. Thermography is a relatively novel method within NDT. It is mainly used for the inspection of ceramics, plastics and composites [11, 12]. Lately, inspection of metal structures and welds are reported [13-16]. Thermography has the advantages that it is relatively fast, non-contacting and provides full field information; therefore it is suitable for an automatic NDT-cell. For readers not familiar with thermography, a short description is presented below. For a more detailed description several references are available [8, 17].

All objects around us with their own temperature above absolute zero emit infrared radiation. Although infrared radiation is invisible to the human eye, it can be converted to a visible image that depicts the variation in temperatures with the help of an IR camera. Thermography is based on this conception. In active thermography the sample is initially excited by a heating device. The redistribution of the temperature on the surface of the sample is then recorded by an IR camera. The thermal images during the temperature-decreasing period are recorded. Cooling devices can also be utilized for generating a temperature difference. Some cooling methods are presenting quite desirable results [18]. With the technique of observing the temperature changes on the surface of the sample after heat excitation, the active thermography test is selected for automatic inspection in this paper. Active thermography is a non-contact and full field method, providing a fast procedure suitable for automation.

For an automatic NDT cell with a thermography system mounted on an industrial robot, the vibration of a robot affects the resulting thermal images from the IR camera, and the results from the thermography test may be disturbed. A non-stable condition is obtained both when the IR camera suddenly stops at one location, and when it moves along the path with a definite speed. In order to determine and evaluate the influence of vibration induced disturbances on the resulting images, this study was initiated. The work is divided into two parts: vibration measurement of the robotic arm and evaluation of the vibration induced disturbances on the results from automatic pulsed thermography tests. By conducting the vibration measurement, the vibration amplitude of a point to point robot movement is measured. To evaluate the influence of the vibration on the thermographic system, the diameters of flat bottom holes, simulating the geometry variation of a weld nugget, were measured during the movements of two different robots at different speeds. In order to study the influence on the resulting IR images, a comparison of the measurement form the thermography system were used as an evaluation method. The results were compared to the result from a measurement without moving the robotic arm. With respect to the quality of the IR-images, the influence of the vibration was evaluated. The remainder of this paper is structured as follows: Section 2 describes how vibration measurement is carried out; Section 3 reviews active thermography and elaborates on the experiments of automatic pulsed thermography; Section 4 presents the results from vibration measurement and the pulsed thermography test and in section 5, conclusion are drawn based on the results.

2   Vibration Measurement

In order to evaluate the influence of the vibration during the robot movement, the vibration on the robotic arm was measured by a mounted IR camera. The vibration measurement was carried out on two types of industrial robots: an ABB robot (IRB 2400) and a STÄUBLI robot (RX-90). Data for the two robots are presented in Table 1. The pulsed thermographic system, including an IR camera and a flash lamp, designed for automatic inspection, was firmly mounted on the robotic arm during the measurement, see Fig. 1.

Table 1. Data of robots compared

Robot Specification	ABB – IRB2400	STÄUBLI – RX-90
Rated payload capacity	12 kg	11 kg
Positional repeatability	±0.03 mm	±0.02 mm
Reachability	1550 mm	985 mm

A three-axis piezo-electric accelerometer (Kistler 5015A) was attached magnetically on the metal plate where the IR camera was mounted. During the vibration measurement, the robotic arm was moving in the x-direction and the vibration amplitude was measured both in x- and y- directions, see Fig. 1. Frequencies in the range of 0-200 Hz were analysed. The vibration amplitude of a point to point robot movement was measured for 5 s including 2 s of robot movement at a speed of 400 mm/s. This made it possible to evaluate the vibration behaviour during the start-up period, the continuous movement and during the settling time.

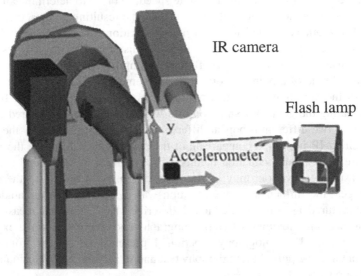

Fig. 1. The robotic arm with the IR camera and the flash lamp mounted in the vibration measurement. A piezo electric accelerometer was attached for measuring the vibration amplitude. The robot was moving in the x-direction during measurement.

3   Active Thermography

In active thermography, the sample is initially excited by a heating device, see Fig. 2, [17]. The redistribution of the temperature on the surface of the sample is then recorded by an IR camera. The thermal images during the temperature decreasing period are recorded. In this paper, pulsed thermography is used. Pulsed thermography uses a pulsed heat excitation, in this case from a flash lamp. Pulsed thermography is a non-contact and full field method, providing a fast procedure suitable for automation. Industrial robots are preferably selected as the scanning equipment because of their flexibility. In an automatic thermography cell, the heating as well as the image-recording process should be triggered automatically at certain points during the test.

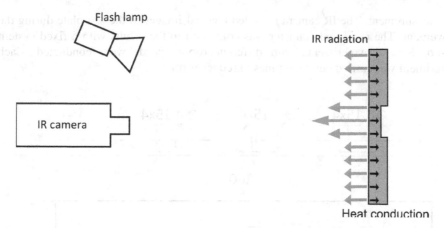

Fig. 2. Principle of pulsed thermography by using IR camera and flash lamp. The sample is initially heated by the flash lamp, the temperature on the surface is recorded by an IR camera.

3.1   Automatic Thermography Test

The influence of the vibration due to the robot movement was analyzed by studying the images from the thermography system on a test plate, simulating the geometry variation of a spot weld. The test plate was made of a 5 mm thick metal plate with three flat bottom holes on the back side. The diameters of the holes were 4.15 mm, 6.15 mm and 8.15 mm, and the depth of the holes was 4 mm. The sizes of the holes were to simulate a range of different sizes of the heat affected zone in spot weld operations on a car body. The test plate is shown in Fig. 3.

In the pulsed thermography test both the flash lamp and IR camera (FLIR SC5650) were mounted on the robot arm and acted on the same side of the test plate where the holes are not visible from the position of the camera, see Fig. 4. The test was started by the robot taking the flash lamp and IR camera to the testing position. An inductive sensor was placed at a certain location in the robot's path. Once the robot moved to the position where the sensor was placed, a signal was sent to the computer as an input to trigger the heating process by the flash lamp. This was followed with image-recording by the IR camera. The flash lamp used gave a short light pulse of 2 ms at 6 kJ and the IR camera used a frame rate of 250 Hz.

The influence of the vibration on the inspection was investigated due to a sudden stop or movement of the robot on the same test plate. Two groups of experiments were conducted to evaluate the influence of vibrations. The first group focused on when the measurement starts right after the robot has positioned the thermography system in the testing position, during the settling time of the robot vibration. All three holes with varying diameter were inspected and the results were compared to the situation with a fixed system set-up, i.e. no movement of the robot. The second investigation was carried out when the robot was moving with constant speed during the measurement. The IR camera recorded thermal images of the test plate during the movement. The measured diameter was compared to the results with a fixed system set-up. Several experiments with different robot speeds were conducted. Each experiment was carried out three times to reduce error.

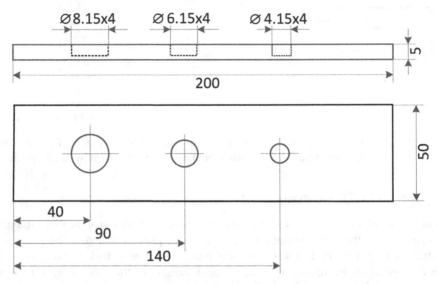

Fig. 3. Test plate used in the thermography test, measures in [mm]

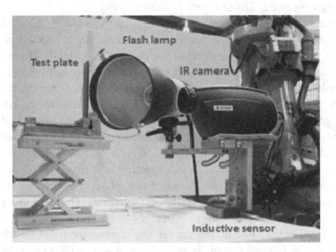

Fig. 4. Experimental set-up for the thermography test

In this paper, a practical technique for quantifying the diameter of the flat bottom hole is provided. Based on the contour-plot image from the thermography system, presenting the cooling rate, a vertical straight line is drawn across the center of the image, see Fig. 5. The line was placed where the largest distance existed in the vertical direction between 2 points at the same temperature. A graph was plotted for the values along the vertical line. By using the two points with maximum gradient as a representation of the edge of the hole, the diameter can be represented by calculating the distance between these two points, see Fig. 5.With this method, the diameter of the hole in the image at any instant can be quantified and followed. The technique will not provide an absolute value of the diameter, but can be used for the purpose of evaluating the influence of the vibration from the robot movement.

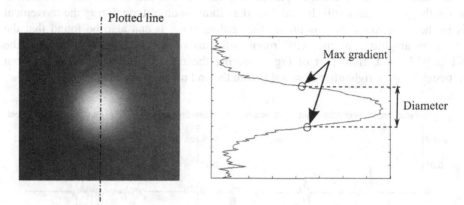

Fig. 5. Measurement of the diameter of nugget in 1st derivative image

Comparing images captured with the IR camera mounted in a fixed position, the position of the holes in adjacent images, captured whit the IR-camera mounted on the moving robotic arm, will vary. To evaluate the diameter with the technique used in this evaluation, different approaches could be used. Instead of determining the vertical line, described above, for every image, a simple technique is suggested. Since the displacement of the holes between two adjacent images is known from the known speed of the robot movement, the translation needed can be calculated. That is how many pixels in the resulting image the robot has moved between two adjacent images. Thermal images are then processed to make sure the holes appear at the same positions in all images. Finally, all the images are stacked back together. After that, the same procedure can be followed to get the measure of the diameter.

4   Results and Discussion

The amplitude of vibration of a point to point robot movement is measured on an ABB IRB 2400 robot and on a STÄUBLI RX-90 robot. The vibration amplitude in the x- and y-direction is presented in Fig. 6. A low pass filter of 200Hz is used in the presented result. As is indicated in the figure, the robot began to move in the x-direction at 1.5 s, and was set to move for 2 s in the robot program. Therefore, the robot stopped at the end location at 3.5 s. The speed of the robot during the 2 s of movement was 400 mm/s

It can be seen from the results in Fig. 6 that the vibration amplitude in the moving direction (the x-direction) is larger than in the perpendicular direction (the y-direction), especially during the start and stop period of the robot. This can be seen by comparing the upper and lower parts of the figure. The highest vibration amplitudes are observed right after the start and before the stop of the robot movement. After the robot stops, during the settling time, there are still some remaining vibrations existing in the system, and it takes the robot about 0.25 s to settle down to the same vibration level it had before any movement began. During the movement of the robot, the speed is constant and the vibration amplitude is quite low, but still noticeable compared to when the robot stands still. In addition, the vibration amplitude during the movement is in the same range as it is during the settling time. It can also be found that the vibration amplitude of the ABB robot, left part of Figure 6, is smaller than the STÄUBLI robot, right part of Fig. 6, during the movement and settling time, but noticeably larger right after start and before the end of the movement.

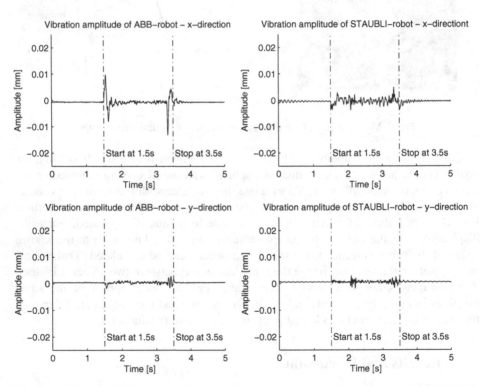

Fig. 6. Amplitude plot for the vibration measurement during the robot movement of ABB (left) and STÄUBLI (right) robots. The amplitude is measured in the moving direction, the x-direction (upper), and perpendicular to the moving direction in the y-direction (lower).

The influence of a sudden stop of the ABB robot in the measurement of the diameter of the flat bottom holes with the pulse thermography system is presented in Fig. 7. All three holes were recorded after the robot finished movement and during the settling time of the robot. The flash lamp in the thermography system heats the test plate shortly after the robot stops, and then the recording by the IR camera starts. As it is shown in Fig. 7, the difference is quite small, less than 2 %, except in the recording of the 4.15 mm hole at 0.05 s after the robot stops, indicating a difference of 6.5 %, which may be explained by the simplified evaluation method used in this study. The measured diameter did vary during the recording, which is due to the procedure chosen for interpreting the data. A better analysis tool is needed for exact measurement of the diameter. In this study, the focus was to compare the measurement with and without vibration disturbances, and the analysis algorithm is not that important as long as the same algorithm is used during the entire study.

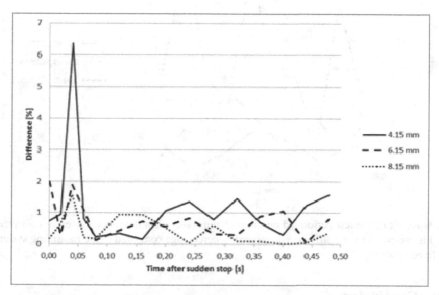

Fig. 7. Result of the pulsed thermography performed on the three different holes of the test plate is presented by difference measured in percentage compared to the measuring system in a fixed position

It is worth mentioning that there is a short delay in the pulsed thermography test between the stop of the robot at the test position and the flash lamp starting to heat the test plate. The delay may somehow explain the results. The robot might have already settled before the testing gets started. This delay is due to the limitation of the method used for triggering the heating process. No matter what kind of method is used, this delay always exists. It can be reduced, but not eliminated.

The influence of the moving robot during measurement is presented in Fig. 8. The 6.15 mm hole was used in the investigation, and the difference between results from a continuously moving robot and a fixed measuring set-up is presented at three different speeds; 20, 40 and 80 mm/s. When the robot is moving with low speed, for example, 20 mm/s or 40 mm/s, not much difference can be observed, less than 5 %, which indicates that the images recorded during the movement are not affected much.

However, at the robot speed of 80 mm/s, the diameter of the nugget presents much larger difference, more than 10 %. The diameter failed to be measured at the speed of 160 mm/s due to the poor image quality due to vibration disturbances, therefore, no data for higher speeds than 80 mm/s is presented here.

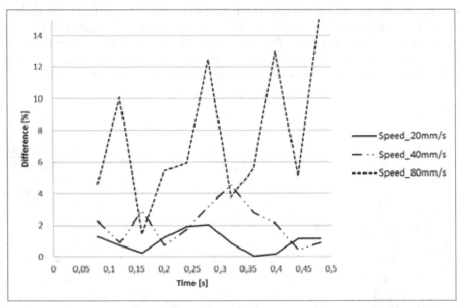

Fig. 8. Result of the test performed on the 6.15 mm hole when the robot moves at a different speed is presented by the difference measured in percentage compared to the measuring system in a fixed position.

5   Conclusion

This paper aims to gain knowledge about the influence of the vibration induced disturbances on automatic thermography test results due to the movement of a robot. The focus is on when the robot stops suddenly at one location, and when it moves at a certain speed. The amplitude of the vibration of the robot movement was measured by a piezo electric accelerometer while the robot was running, at a speed of 400 mm/s. It was found that the vibration amplitude differs between different robots. One has to consider the choice of robot when developing an automatic thermography inspection system. It should be possible to calculate a correlation constant between different manipulators, in order to quantify them. This is anyhow not the scope of this paper. The vibration amplitude also varied during the movement with larger vibration amplitude during the start and stop period. During the settling time, the vibration amplitude was almost in the same range as during the continuous movement.

A thermography system was mounted on the ABB robot, the robot with the lowest vibration amplitude found in the prior experiment, for the inspection of automatic pulsed thermography. In the images recorded after the robot stopped moving, the influence on the diameter is small. From the result, the error percentage is not dependent on the size of the hole observed. Due to the simplified analysis tool used in this study, no absolute conclusion of the influence of the vibration on the measured diameter could be drawn, but a comparison between the different experiments is possible. The influence of vibration when the robot was moving at certain speeds varies by the running speed of the robot. When the robot was moving at a speed of 20 mm/s or 40 mm/s, the influence is found to be quite low, less than 5 %. However, when the robot moves at a speed of 80 mm/s, the influence of vibration is larger and not in the same range compared to lower speed. The speed of the robot was significantly lower during the thermography experiment than during the accelerometer measurement, even though the influence on the thermographic measurement, during the continuous movement, was significant, starting at 80 mm/s. Therefore the resolution of the resulting images will be more important as the speed increases. This might be due to an operation deflection shape, excited at the speed, but could also be due to influence from the servo engines. This has to be studied further. From this study, it is concluded that the influence of the vibration should be taken into consideration, thus vibration needs to be controlled when the robot moves at certain speeds in an automatic inspection cell.

References

1. Carvalho, A.A., et al.: Reliability of non-destructive test techniques in the inspection of pipelines used in the oil industry. International Journal of Pressure Vessels and Piping 85, 745–751 (2008)
2. Runnemalm, A.: Vibration Induced Disturbances in Automatic Non-destructive Test. Presented at the 18th World Conference on Nondestructive Testing, Durban, South Africa (2012)
3. Nilsson, P., et al.: Automatic Ultrasonic testing for Metal Deposition. Presented at the 18th World Conference on Nondestructive Testing, Durban, South Africa (2012)
4. Ismail, A.R., et al.: The performance analysis of industrial robot under loaded conditions and various distance. International Journal on Mathematical Models And Methods In Applied Sciences, 277–284 (2008)
5. Haitao, L., et al.: Vibration characteristic analysis of spot-welding robot based on ADAMS/Vibration. In: 2011 International Conference on Electronic and Mechanical Engineering and Information Technology (EMEIT), pp. 691–695 (2011)
6. Juyi, P., et al.: Design of learning input shaping technique for residual vibration suppression in an industrial robot. IEEE/ASME Transactions on Mechatronics 11, 55–65 (2006)
7. Ohishi, K.: Robust position servo system based on vibration suppression control for industrial robotics. In: Power Electronics Conference (IPEC), pp. 2230–2237 (2010)
8. Shull, P.J.: Nondestructive evaluation: theory, techniques and applications. Marcel Dekker, New York (2002)
9. Åström, T.: From Fifteen to Two Hundred NDT- methods in Fifty Years. In: Presented at the 17th World Conference on Nondestructive Testing, Shanghai, China (2008)
10. Raj, B., et al.: Non-destructive testing of welds Pangbourne: Alpha Science (2000)

11. Hamzah, A.R., et al.: An experimental investigation of defect sizing by transient thermography. Insight: Non-Destructive Testing and Condition Monitoring 38, 167–173 (1996)
12. Hung, Y.Y., et al.: Review and comparison of shearography and active thermography for nondestructive evaluation. Materials Science and Engineering: R: Reports 64, 73–112 (2009)
13. Almond, D.P., et al.: Thermographic techniques for the detection of cracks in metallic components. Insight: Non-Destructive Testing and Condition Monitoring 53, 614–620 (2011)
14. Lee, S., et al.: A study on integrity assessment of the resistance spot weld by Infrared Thermography. Procedia Engineering 10, 1748–1753 (2011)
15. Meola, C., et al.: The use of infrared thermography for nondestructive evaluation of joints. Infrared Physics and Technology 46, 93–99 (2004)
16. Schlichting, J., et al.: Thermographic testing of spot welds. NDT & E International 48, 23–29 (2012)
17. Maldague, X.P.V.: Theory and practice of infrared technology for nondestructive testing. Wiley, New York (2001)
18. Woo, W., et al.: Application of infrared imaging for quality inspection in resistance spot welds. In: Proceedings of SPIE - The International Society for Optical Engineering, Orlando, FL (2009)

Part Alignment Identification
and Adaptive Pick-and-Place Operation for Flat Surfaces

Paulo Moreira da Costa[1], Paulo Costa[1,2], Pedro Costa[1,2], José Lima[1,3],
and Germano Veiga[1]

[1] INESC TEC (formerly INESC Porto)
[2] Faculty of Engineering, University of Porto
[3] School of Engineering, Polytechnic Institute of Bragança
paulojorgemcosta@gmail.com, {paulo.j.costa,pedro.g.costa,
jose.lima,germano.veiga}@inescporto.pt

Abstract. Industrial laser cutting machines use a type of support base that sometimes causes the cut metal parts to tilt or fall, which hinders the robot from picking the parts after cutting. The objective of this work is to calculate the 3D orientation of these metal parts with relation to the main metal sheet to successfully perform the subsequent robotic pick-and-place operation. For the perception part the system relies on the low cost 3D sensing Microsoft Kinect, which is responsible for mapping the environment. The previously known part positions are mapped in the new environment and then a plane fitting algorithm is applied to obtain its 3D orientation. The implemented algorithm is able to detect if the piece has fallen or not. If not, the algorithm calculates the orientation of each piece separately. This information is later used for the robot manipulator to perform the pick-and-place operation with the correct tool orientation. This makes it possible to automate a manufacturing process that is entirely human dependent nowadays.

Keywords: kinect, 3d vision , Pick-and-place, robotic manipulator.

1   Introduction

Laser cutting machines are widely used on metallurgical industry. Even though there are different manufactures, the machines share the same basic kinematic structure, which consists of a Cartesian robot coupled with a laser that covers the entire workspace. The main metal sheet lies on a metal support base that maximizes the presence of air below the metal sheet to be cut. This is ensured using a support composed of vertical triangles where the metal lies only on its tips, Figure 1a).

With this architecture the metal parts tend to tilt or fall after the cut, making the robot's collection task more complex, Figure 1b). In order to perform the pick-and-place operation, the robot needs to perceive the misalignment of the cut parts to enable the automation of the subsequent picking operation. The use of an industrial robot for this operation requires object pose identification because the piece extraction trajectory depends on its orientation at that time.

P. Neto and A.P. Moreira (Eds.): WRSM 2013, CCIS 371, pp. 203–212, 2013.

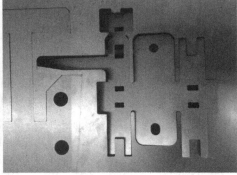

a) Laser cutting machine [1]   b) Tilted piece after cut

Fig. 1. Laser cutting machine and cut metal piece example

This paper is divided in five chapters including the introduction. In the next chapter the state of the art is presented. The process to fulfil the objectives is described in chapter three. In chapter four the results are presented and discussed. Finally, in chapter five the paper is concluded and future work is proposed.

1.1   Objectives

The objective of this work consists of detecting the 3D position and orientation of cut metal parts in order to successfully perform the pick-and-place operation. This means that the robot has to decide if the pieces can be picked or not. In the affirmative case, the robot has to approach the metal part with correct tool orientation angle. In other words, the system has to adaptively perform the pick-and-place operation with regard to the piece to be collected, always avoiding picking the fallen ones.

2   State of the Art

Similar work can be found on bin picking related projects that include research on perception, grasping and path planning algorithms. Perception is the most relevant aspect in this work and, therefore, the state of the art presented here will focus only on that area. This is because the metal parts have a flat contact area that facilitates grasping techniques. The path planning is also simplified since there are no obstacles during the pick-and-place operation. [2, 3]

The system relies on three-dimension vision hardware. These technologies can be active or passive depending on whether there is interaction with the environment or not, respectively. Due to their mode of operation and sensor characteristics these technologies can be divided as (a) triangulation based active ranging (PrimeSense technologies), (b) vision based passive ranging (stereoscopic vision), and (c) time-of-flight active ranging (laser rangefinders). Triangulation based active ranging technologies use geometric properties manifested in their measuring strategy to

Fig. 2. Microsft Kinect, developed by PrimeSense

establish distance readings to objects. Vision based passive ranging technologies are sensing devices that capture the same raw information light that the human vision system uses. Finally, time-of-flight active ranging technologies makes use of the propagation speed of sound or an electromagnetic wave. [4, 5]

PrimeSense [6] is responsible for developing the Microsoft Kinect [7], shown in Figure 2, and the Asus Xtion [8]. They share the same work principles and they are known for their good performance at a considerably low price. In addition, the work requirement ranges fit the technical limitations of the Microsoft Kinect. Internally, the Kinect contains an RGB camera, an infrared (IR) camera and an IR projector. Its three-dimensional vision characteristics come from triangulation between two consecutive IR frames. It is possible to build a colorized point cloud by mapping the depth map with the information from the RGB camera. There is a wide range of research groups developing computer vision solutions based on this technology. All the facts considered make this hardware suitable for the perception subsystem implementation [9, 10, 11].

3   Methodology

The implemented system depends on perception and robot control. The perception component relies exclusively on the Microsoft Kinect. The image data is later processed together with the known pieces position and format information used as input data for the laser cutting machine side. This, combined with a plane fitting algorithm, returns the pick position and orientation that serves as input for the robot trajectory control.

3.1   Architecture

Due to the industrial nature of the project, it was necessary to simulate the working environment in the laboratory. The hardware architecture implemented is mainly divided as follows: Microsoft Kinect, ABB Robot, computer, cut metal sheet provided by a company in this field, and a wooden prototype support base, Figure 3.

A high level application is responsible for controlling the perception hardware and the data is shared with the robot via serial communication. The software solves the computer vision algorithms and uploads to the robot the calculated position and pose of the robot target.

a) Laboratory architecture   b) Main metal sheet

Fig. 3. Set of hardware used for the system implementation

3.2   Perception

The perception is responsible for solving the sensing raw data coming from specific algorithms that provide meaning to the acquired environment information. The Kinect observes the environment and builds a depth map with pixel values proportional to the object distance. To turn the raw calculated distances into SI units, the conversion method presented in Equation 1 is used, as proposed in [12].

$$d_m = 0.1236 \tan(\frac{d_k}{2842.5} + 1.1863) \tag{1}$$

where dk is the raw depth to a specific point directly provided by the Kinect; dm represents its conversion to meters.

Considering this dimension as z axis on the Kinect reference frame, then x and y are defined according to their width and height. These last two dimensions need to be interpolated from the depth distances taken from the Kinect, as show in Equation 2.

$$P_c(u,v) = \begin{bmatrix} x_c(u,v) \\ y_c(u,v) \\ z_c(u,v) \end{bmatrix} = d_m(u,v) \times \begin{bmatrix} (u - c_x^{IR})/f_x^{IR} \\ (v - c_y^{IR})/f_y^{IR} \\ 1 \end{bmatrix} \tag{2}$$

where, Pc are the coordinates in meters related to the IR camera frame; u and v are the coordinates in pixels also related to the IR camera frame; f and c are the intrinsic parameters of the IR camera (f focal length and c the distance between the lens and the focal point).

At this point, the depth map returned from the computer vision hardware contains all the pixels mapped to the camera reference frame in meters. However, it is useful to have the depth map referenced to a frame that is shared with the robot so that a specific 3D point has the same definition both for the Kinect and robot frames, that is, the world reference frame. This results in a homogeneous transformation from the camera to world reference frames, Equation 3.

$$P_w = (P_c - T) \times R \tag{3}$$

where, Pc are the coordinates in camera reference frame, and R and T are the rotation and translation matrices, respectively.

In this project, the Kabsch algorithm [13] was used in order to calculate the matrices responsible for the above mentioned transformation. This algorithm uses two sets of paired points, where one is referenced to the Kinect and the other is referenced to the world frame. Firstly, the translation is calculated by taking the centroids of the two meshes and the consequent distance between them. Both sets of points are then centred on their respective centroids. Secondly, it uses the covariance matrix to calculate the optimal rotation matrix that minimizes the root mean squared deviation error between both sets. From this point, the depth map is referenced to the same robot work frame. As a result, it is possible for the robot to work directly with the coordinates returned from the Kinect.

The Kinect acquires more than the region of interest. Therefore, after the overall environment mapping, it is advantageous to work only in the region of interest to achieve faster processing times. After transforming the coordinates, this is easily done by disregarding the points outside a specific range of values for the three dimensions in x, y and z in the world frame.

This computer vision hardware returns null reading values for points where it was not possible to calculate the distance. Since the current frame depends on the previous frame, a reading error on the previous frame affects the accuracy of the current frame for that specific point. Consequently, a simple pre-processing technique is applied in order to increase the quality of the depth map. This technique consists of calculating the median of three consecutive frames where the median depth is only calculated for 3D points where no reading errors occurred. This increases the number of null values; however, it increases the reliability of the depth map. An example of this method is presented in Figure 4.

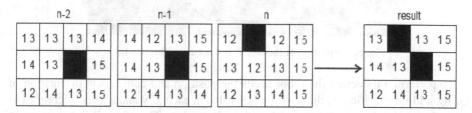

Fig. 4. Illustration of the pre-processing technique

3.3   Pose Identification

In order to calculate the orientation angle of a specific metal part, the algorithm implemented starts by matching the known piece positions to the depth map returned from the computer vision hardware. Therefore, there is no implicit piece detection based on image processing algorithms. Alternatively, since both the Kinect and the robot have the same reference frame, the known positions of the metal parts can be mapped directly onto the depth map. For the plane fitting algorithm, only a set of points is considered that match a circle whose radius is proportional to the size of the metal part. Thus, the algorithm uses a limited set of values based on one point from the previously known data, originated from the laser cutting machine design software. The mentioned group of points maps a delimited circular region for each piece that works as input for the plane fitting algorithm. Its implementation uses singular value decomposition (SVD) and returns the normal vector to the plane defined from the input points referenced to its orthonormal reference frame. With this normal it is possible to calculate the orientation magnitude between this vector and the normal of the main metal sheet. If the world reference frame has xOy coinciding with the metal sheet plane, then its normal will have the direction of z. The magnitude orientation is solved as an ordinary angle calculation between the vectors zw and zn, where w and n represent the world and piece (based on the normal vector) reference frames, respectively, Figure 5. The returned normal vector contains more relevant information since its projection in the world's xOy plane reveals the 2D orientation of the tilted piece.

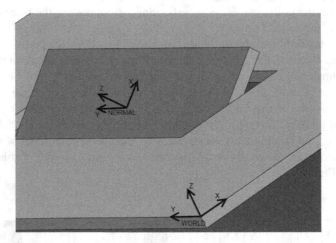

Fig. 5. Illustration of the world and normal reference frames

To perform a trajectory the robot needs the position and consequent orientation of the tool. More specifically, the position is a value in each x, y and z axes referenced to some frame and the orientation is set with a quaternion. Therefore, it is necessary to have an orthonormal reference frame in each metal sheet where both the position and orientation are mapped. This is done by applying the plane fitting algorithm knowing that the world reference frame is already set in XYZw. Then it is possible to calculate the other two axes that together with the zn build an orthonormal reference frame XYZn, Equation 4 and 5.

$$y_n = z_n \times x_w \tag{4}$$

$$x_n = y_n \times z_n \tag{5}$$

With these simple cross product calculations, and using the normal vector returned from the plane fitting algorithm, it is possible to map both the position and orientation angle using an orthonormal reference frame. The frame origin maps the position in the world reference frame, and the orientation is provided by the deviation between both reference frame axes, Figure 5.

3.4   Robot Control

At this stage the perception algorithm provides all the input data necessary for the robot to perform the pick-and-place trajectory, that is, the calibrated world reference frame, and the position and orientation of the metal part. The pick positions with correct orientation are the robot targets in a specific robot trajectory.

The system contains all the input data necessary for the robot to perform the pick-and-place trajectory. The Kinect and the robot share the same work reference frame and the perception system is able to calculate both the position and orientation for each metal piece.

Therefore, the robot is controlled using the previously known position and calculated orientation as input data. This data is transferred from the industrial computer to the robot over serial communication. The robot receives the data and computes it iteratively for each metal part.

4   Results

The results will cover the two main parts of this project: perception and robot picking performance, using the previously presented architecture. Firstly, examples of the perception algorithm are presented and the results are discussed. Secondly, a number of consecutive picking operations are performed in order to numerically approximate the robot picking reliability. The tests consist of putting the cut pieces aligned with the main metal sheet and letting them rearrange arbitrarily. This simulates a normal scenario where the main cut metal parts come from the laser with unknown orientations.

Figure 6 shows two images acquired with the Kinect, both representing the same scenario directly seen from its point of view: colorized scene from the RGB camera, Figure 6a), and depth map in grey scale with circular regions of interest for the plane fitting algorithm, Figure 6b). In the depth map, darker colours mean farther distances to the Kinect and the black colour maps points outside of the area of work or with unknown distances. In the same picture is evident the surrounded main metal sheet area which represents the area of work. This area is automatically obtained after the world reference frame calibration. The perception software classifies the pieces according to their orientation: green means alignment with the metal sheet (no orientation), orange (tilted piece) and red means invalid orientation (absent or fallen piece).

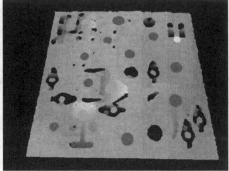

a) RGB frame   b) Depth in grey scale with circular regions of
                                           interest

Fig. 6. View from the Kinect of an example scenario that includes tilted and fallen pieces

For the misaligned pieces, it was possible to compare the plane fitting results to the measurements taken from the piece itself regarding its orientation relatively to the main metal sheet. The plane fitting calculations were conducted three consecutive times to make it possible to study the repeatability performance of the algorithm. The averages of these calculations can then be compared to the measurements. This test was performed on four different pieces at increasing distances from the Kinect. This means that piece 1 is the closest (1,0m) and piece 4 is the farthest (1,70m). The results are presented in Table 1. In the last column, the standard deviation of the three plane fitting calculations shows that the repeatability decreases as the distance from the objects increases. The plane fitting calculation error is also consistent with the Kinect error dynamic because it increases proportionally to the distance. These numbers show that the distance affects both the repeatability and the accuracy as the algorithm depends directly on the performance of the Kinect. Piece 4 represents the farthest piece on the work area and, therefore, it approximates the highest error for the plane fitting algorithm. This accuracy is sufficient for the system validation, because collecting the piece with magnetic or vacuum tool has some orientation compliance. Therefore, this small error (only evident for longer distances) does not jeopardize the picking operation.

Table 1. Comparison between measurements and plane fitting calculations

Pieces	Measurements	Plane fitting calculations			μ	μ-M	σ
1	20°	19°	19°	20°	19.3°	0.7°	0.6
2	25°	25°	26°	27°	26.0°	1.0°	1.0
3	22°	25°	27°	26°	26.0°	4.0°	1.0
4	21°	26°	27°	23°	25.3°	4.3°	2.1

The robot was coupled with a magnetic gripper, as demonstrated in Figure 7, in order to test the overall system in the laboratory test-bed. The geometry of the metal parts hinders the pick of the tilted pieces since they get stuck in the process. It is impossible for the robot to know this information a priori. The pieces are explicitly classified as aligned if they present an absolute orientation angle below five degrees, represented in green in Figure 6b). For these cases, the robot was able to successfully pick all pieces for three consecutive times in its work range without any failure. For the cases where picking is impossible, the robot was also able to successfully align with all of them, as demonstrated in Figure 7.

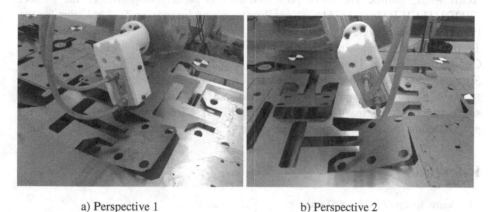

a) Perspective 1   b) Perspective 2

Fig. 7. Tool approach with magnetic gripper for tilted piece

5   Conclusion

The chosen computer vision hardware presented good performance results when calculating the depth map. The implemented conversion to SI units, associated with reference frame calibration, made it possible to easily share the results from perception hardware with the robot. The plane fitting algorithm returned accurate results for the normal vector which is accurate enough for the problem considered.

This implementation shows that low cost vision hardware such as the Kinect can be used for industrial applications. The precision is sufficient even when working on its technical limitations. The results are excellent when working for closer distances. As a consequence, the position of the Kinect should be previously studied to take advantage of its best performance.

Finally, the robot is able to perform the pick-and-place operation using the information from the perception subsystem. With the result from the plane fitting algorithm the robot can decide whether to pick, to approach or to avoid a specific metal piece. The picking of the aligned pieces demonstrated an excellent performance and the approach is also very accurate with piece orientation.

5.1   Future Work

The system implemented uses the Kinect, which has a limited area of work. It would be interesting to upgrade the system to work with multiple Kinect systems or similar sensors. This would increase the area of work and also the quality of the depth when overlapping the information obtained with multiple sensors. If the sensors are positioned correctly, it is possible to avoid occlusions, thus significantly reducing null data.

Acknowledgements. The work presented in this paper, being part of the Project PRODUTECH PTI (nº 13851) – New Processes and Innovative Technologies for the Production Technologies Industry, has been partly funded by the Incentive System for Technology Research and Development in Companies (SI I&DT), under the Competitive Factors Thematic Operational Programme, of the Portuguese National Strategic Reference Framework, and EU's European Regional Development Fund".

The authors also thanks the FCT (Fundação para a Ciência e Tecnologia) for supporting this work trough the project PTDC/EME-CRO/114595/2009 - High-Level programming for industrial robotic cells: capturing human body motion.

References

1. Adira, http://www.adira.pt
2. Song, K.-T., Tsai, S.: Vision-based adaptive grasping of a humanoid robot arm. 2012 IEEE International Conference on Automation and Logistics (ICAL), 155–160, 15–17 (2012)
3. Pinto, M., Moreira, A.P., Costa, P., Ferreira, M., Malheiros, P.: Robotic manipulator and artificial vision system for picking cork pieces in a conveyor belt. In: 10th Conference on Mobile Robots and Competitions, Robotica (2010)
4. Siegwart, R., Nourbakhsh, I.: Introduction to Autonomous Mobile Robots. Bradford Company, Scituate (2004)
5. Moreira da Costa, P.: Operação de "Pick-and-place" Adaptativo para Ambientes Pouco Estruturados. Master Thesis (2012)
6. PrimeSense official site, http://www.primesense.com
7. Microsoft. Xbox 360 kinect, http://www.xbox.com/kinect/
8. Asus Xtion,
   http://www.asus.com/Multimedia/Motion_Sensor/Xtion_PRO/
9. Technical description of Kinect calibration,
   http://www.ros.org/wiki/kinect_calibration/technical
10. Khoshelham, K., Elberink, S.: Accuracy and Resolution of Kinect Depth Data for Indoor Mapping Applications. In: Proceedings of Sensors 2012, pp. 1437–1454 (2012)
11. Khoshelham, K.: Accuracy Analysis of Kinect Depth Data. Int. Arch. Photogramm. Remote Sens. Spatial Inf. Sci., XXXVIII-5/W12, 133–138 (2011)
12. Image information, http://openkinect.org/wiki/Imaging_Information
13. Kabsch, W.: Automatic processing of rotation diffraction data from crystals of initially unknown symmetry and cell constants. J. Appl. Cryst. 26, 795–800 (1993)

Manipulator Path Planning for Pick-and-Place Operations with Obstacles Avoidance: An A* Algorithm Approach

João Sousa e Silva[1], Pedro Costa[1], and José Lima[2]

[1] INESC TEC (formerly INESC Porto) and Faculty of Engineering,
University of Porto, Portugal
{joao.r.silva,pedro.g.costa}@inescporto.pt
[2] INESC TEC (formerly INESC Porto) and Polytechnic Institute of Bragança, Portugal
jllima@ipb.pt

Abstract. This paper presents a path planning method for pick-and-place operations with obstacles in the work environment. The method developed is designed to plan the motion of an anthropomorphic manipulator in cluttered environments. The graph search algorithm A* applied to the configuration free space is used to calculate the shortest path between two points avoiding collisions with obstacles and joint limitations. Applying this algorithm in a six dimension space presents some constraints related to memory consumption and processing time, which were tackled using configuration space partition and selecting neighbourhood cells, respectively. Using the configuration space makes it possible for the entire robot body to avoid collisions with obstacles. Moreover, the system implemented proves that applying A* in high dimension configuration spaces is possible with admissible results.

Keywords: path planning, obstacle avoidance, A*, anthropomorphic manipulator.

1   Introduction

In many robotic applications, the existing obstacles in work environments may hinder the robot's actions when performing a task. This work presents a pick-and-place application in which the robot motion is constrained by the presence of obstacles. In this particular case, there is a single obstacle – a conveyor belt used to transport objects to be collected. Therefore, it is necessary for the robot to perform its task successfully without colliding with the conveyor. In order to achieve that, motion planning is essential for the manipulator performance operation. Nevertheless, the problem focuses on avoiding collision of the entire robot body with the obstacle, not only with the tool central point. Thus, during the execution of the task, it is necessary to perform a configuration planning to be assumed by the manipulator. This is

P. Neto and A.P. Moreira (Eds.): WRSM 2013, CCIS 371, pp. 213–224, 2013.
© Springer-Verlag Berlin Heidelberg 2013

considered because the positioning of the tool central point in the free space does not ensure that another point of the robot is not in contact with the obstacle. Section 2 of this paper presents a state of the art in domain of path planning systems. The methodology following the implementation of the planning system is described in Section 3. Section 4 presents the results of system implementation. Finally, Section 5 describes the conclusions obtained and future work to be developed.

1.1   Objectives

The objective of this work is to develop a path planning system. In the presence of obstacles, path planning consists of finding paths free of collisions that connect two previously defined configurations. The implemented system should be able to compute paths considering a configuration space approach. The dimension of this space is equal to the number of manipulator joints and it is possible to define a configuration by a single point. In configuration space, the problem can be simplified by searching paths free of collisions between two points that belong to the space. This type of planning does not take into account aspects related to dynamic of motion, such as the speed or the acceleration of manipulator joints to execute the desired movement.

2   State of the Art

The classic path planning algorithms can be classified into three types: *roadmap, cell decomposition* and *potential fields* [1].

In the roadmap method, a roadmap is a way of representing the free configuration space and its connectivity. It is composed of a network of free paths that the robot is able to execute. A well-known roadmap method was proposed by Canny [2] - the Silhouette method. With this method semi-free paths are obtained in the boundary of free space configuration. The decomposition cell method divides the space to be searched into a set of cells which are represented by a connectivity graph [1]. A graph search algorithm is used to find a path between two cells belonging to this space. Two types of decomposition can be identified: *exact cell decomposition* and *approximated cell decomposition*. Wu and Hori [3] present a real-time path planning algorithm where the environment is represented using the *Octree* decomposition. *Octree* is an approximated cell decomposition method where the cells are consecutively subdivided until there are no mixed cells in the map. Furthermore, Wu and Hori have shown an application of potential field algorithms based on information of the *Octree* model, where the field is given by a function of the level node on the *Octree*. With the potential field algorithm, the manipulator motion is defined by the action of an artificial potential field. The manipulator is repelled by obstacles and attracted by the goal configuration.

In addition to previous classic methods, other types of algorithms can be used to solve the path planning problem. These methods are called probabilistic and aim to obtain free paths between configurations using random approaches. Recently, this type of methodology has been studied to be applied in systems with a high degree of freedom. This happens because a full representation of the space where the search of a path will be made is not essential. There are two widely studied algorithms which fall into this type of method: the *Probabilistic Roadmap (PRM)* and the *Rapidly-Exploring Random Tree (RRT)*. The first method was proposed by Kravaki et al. [4] and it is a particular case of the previously presented roadmap method. However, in this case the roadmap is built incrementally using probabilistic techniques. The RRT is an algorithm with a quick execution that can be used in real-time applications and was introduced by LaValle [5]. The tree is expanded along the configuration space, starting on the initial node until the target node is found. A path planning application of this method is presented in [6]. The last efforts in the path planning domain focus on operations where it is necessary for a robot to cooperate with a human or with another robot [7][8]. In this scope, Costa e Silva et al [9] present a model for generating trajectories of a high degree of freedom manipulator, considering human-like abilities in their cognitive and motor behaviour. For the authors, these capabilities are needed to enable the robot operation in human environments. An important feature of the systems that perform this type of tasks lies on their implementation in dynamic work environments, where a constant perception of the object's position in the space is required.

3   Methodology

Three stages are considered in order to develop the planning system: in the first stage, the configuration space is defined; in the second, the planning method to be used is developed; and in the last stage the graph search algorithm is implemented. Additionally, a method used to send reference values to the manipulator joint controller is presented.

3.1   Simulation Model

A model of a serial manipulator using the SimTwo [10] simulation software was developed to test and validate the implemented algorithms. The software SimTwo is a physics-based simulator which uses ODE (Open Dynamics Engine), an open source library for simulating rigid body dynamics [11][12]. The SimTwo allows the simulation of different types of robots and allows the access to several variables that define the physical and dynamic behaviour of the developed model [11]. A comparison between the dynamic behaviour of a real humanoid robot and simulation model built with this software is done in [11]. The results presented by the authors shown that simulator behaves as the real robot.

In order to create a model as reliable as possible, the dimensional and mechanical characteristics of a real manipulator that is presently commercialized were used. The

Fig. 1. Simulation model used to test the path planning system

robot chosen was the IRB140 from the ABB manufacturer. In addition to the manipulator, a conveyor belt was modelled that is an obstacle placed in the work environment. This simulation model is presented in Figure 1.

3.2   Configuration Space

In the path planning domain, a configuration consists of a complete specification of the location of every point on the robot. The configuration space C is defined by the set of all possible configurations that can be assumed by the robot. Let $q = (\theta_1, \theta_2, ..., \theta_6)^T$ be a configuration where the parameters represent the angular positions of each revolute joint. Considering forward kinematics, it is possible to specify the position of any point on the robot from the value of q.

The configuration space is composed of two subspaces. The first one is called free space C_{free} since for the configurations that belong to this subspace there are no collisions with obstacles. The other one is the obstacle configuration space $C_{obstacle}$ where the opposite occurs. This work uses a static configuration space where the obstacle is a conveyor belt. The approach to compute the configuration space associated to the work environment is based on boundary equation of obstacles. These equations are obtained from trigonometric analysis of the robot's configuration regarding the obstacle. For a 6-DOF manipulator, this is an operation with a high level of complexity and for this reason the links that constitute the spherical wrist are not considered. Therefore, it is necessary to ensure that spherical wrist never collides with obstacles. In order to achieve that, the obstacles are expanded according to the radius of the wrist in the work space. Additionally, configurations which are unreachable due to the mechanical constraints of robot joints belong to $C_{obstacle}$. Thus, boundary equations were calculated considering the schematics shown in Figure 2.

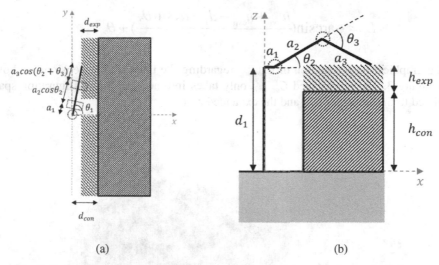

Fig. 2. Conveyor top view (a) and side view (b) for boundary equations calculation

Considering variable θ_1 and fixed θ_2 and θ_3 , it is possible to define three different ranges for value of θ_1:

- From the negative mechanical limit of joint 1 to the value corresponding to the first intersection of the manipulator with the plane that contains the lateral of conveyor, $-L_1$;

- The second range contains the values of θ_1 for which the manipulator is in contact with the conveyor, $]-L_1; L_1[$;

- The final interval represents the values of θ_1 from the last point of contact with the conveyor side plane L_1, to the positive mechanical limit of joint 1.

Moreover, it is possible to verify that L_1 depends on the value of θ_2 and θ_3. It can be calculated from limit case of Figure 2a, by equation (1).

$$L_1 = \arccos(\frac{d_{con} - d_{exp}}{a_1 + a_2 \cos\theta_2 + a_3 \cos(\theta_2 + \theta_3)}) \qquad (1)$$

With the setup used, in the worst case scenario a collision occurs between link a_2 and the top of the obstacle ($\theta_2 = -10°$). However, in practice this collision never happens because it only occurs with the volume that was expanded. Once the spherical wrist is attached to the end of link a_3, there is not risk of a real collision between link a_2 and the top of the conveyor.

The range of values of θ_3 that belong to C_{free} can be calculated from Figure 2b. Considering that the manipulator only collides with obstacles for $\theta_1 \in]-L_1; L_1 [$, the lower limit value for θ_3 is given by equation (2).

$$L_3 = -\arcsin(\frac{h_{con} + h_{exp} - d_1 - a_2\sin\theta_2}{a_3}) + \theta_2 \tag{2}$$

The graphical representation of $C_{obstacle}$ regarding the three first dimensions is shown in Figure 3. The represented $C_{obstacle}$ only takes into account the configuration space related to the conveyor belt and the expanded zone.

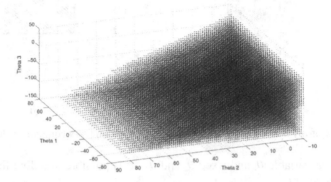

Fig. 3. Representation of obstacle configuration space $C_{obstacle}$ (in degrees)

3.3 Planning Method

The planning method develops an approach based on approximate cell decomposition. Due to several problems related to its application on high dimension spaces, some issues must be taken into consideration. These difficulties have to do with large amounts of data used to represent configuration space maps.

The first procedure involves the partial representation of C, since it was verified that the required memory for a total representation exceeds the free memory available. With this methodology, the map is initialized with a representation of a region of C by default. If the final configuration q_f is in a mapped zone, then the current representation remains valid. On the other hand, if q_f is out of the limits of the referred region, then the map saved in memory is updated. In this case, the limits of the map are defined considering initial and final configurations. The resulting map is held while it is not necessary to reach positions that are not contained in the map. The map limits are defined taking into account two cells of separation between the limits and the initial and final configurations. This is performed so as not to restrict the action of graph search algorithm, allowing approximations to the final configuration from all directions.

However, as the map size is directly proportional to the distance between configurations, the risk of exceeding the free memory available is still present. Therefore, to avoid this problem, maps are created with a number of fixed cells for each dimension and, consequently, the memory consumed is the same for every calculated map. This procedure leads to a variable cell size and also to a variable resolution of the map. For these reasons, the number of cells was chosen considering two criteria: the volume of data produced for representation and the accuracy of the map, considering the size of cells that compose it. The increase in resolution means that more data needs to be saved and processed. Instead, the decrease in resolution may result in a path composed of sparse configurations and, consequently, the movements of the manipulator may be unstable.

The implementation does not explicitly represent the map that is calculated because an individual evaluation is performed for each configuration, verifying whether or not it belongs to $C_{obstacle}$. This evaluation is performed during the execution of the graph search algorithm, which makes it possible to save processing time spent in computing the entire $C_{obstacle}$. The approach used only considers the evaluation of points that have an influence on the calculated path.

3.4 Graph Search Algorithm

As a result of the decomposition cell method, it is possible to obtain a graph that represents adjacency between cells of C. The A* search algorithm is used to find a path free of collisions along this graph. This algorithm uses an evaluation function to estimate the total cost from the initial node n_i to the final node n_f when node n is being exploited. It can be described by expression (3). A complete description of the A* algorithm is provided in [13].

$$f(n) = g(n) + h(n) \tag{3}$$

Where $g(n)$ represents the path cost from n_i to n, $h(n)$ is a heuristic function used to estimate the cost from n to n_f. The function $f(n)$ is used in evaluation of the next node to be exploited by the algorithm. The heuristic function adopted is based on the Euclidean norm. Thus, the straight distance between node $n=(n_1,n_2,...,n_6)$ and node n_f $=(n_{f1},n_{f2},...,n_{f6})$ is calculated as shown in equation (4).

$$h(n) = \sqrt{(n_1 - n_{f1})^2 + (n_2 - n_{f2})^2 + ... + (n_6 - n_{f6})^2} \tag{4}$$

The number of contiguous cells visited at each iteration is given by 3^m-1, where m represents the dimension of the searched space. In this particular case, 728 neighbour cells are visited. For each cell, it is verified if corresponding configuration leads to a collision with an obstacle. If this occurs, the cell is ignored and it is not processed, and the processing of the remaining cells in the neighbourhood is resumed. In addition to cells belonging to $C_{obcstacle}$, cells on the boundary of C are also considered obstacles. This ensures that the algorithm does not visit cells in non-mapped spaces and, consequently, does not return paths that cross this zone.

3.5 Feedback to Set Reference Values

After the path is calculated, reference values are sent to the manipulator in order to execute the desired motion. The used procedure is composed of three stages: receiving information concerning the current configuration of the manipulator;

executing the path planning methodology; and sending new reference values for each joint, considering the free path obtained in the previous step. Considering that the manipulator receives information cyclically, the path is always calculated from the present configuration to the final desired configuration. This planning is performed each time new information related to the robot configuration is received. The cycle time between the consecutive reception of current configurations is defined by the simulation cycle and it is equal to 40 *ms*. The procedure used to send new reference values is initiated by selecting one of the configurations in the set defined by the path planning. The decision related to this choice is motivated by two factors that negatively influence the manipulator's behaviour. The selection of references near current positions causes a shortest distance to go through. The joint controller begins to decelerate when the joint position gets close to the reference value. Therefore, an irregular motion composed of successive accelerations and decelerations of joints would be produced. To avoid this problem, a more distant reference value is chosen so that the next reference can be sent before the deceleration stage.

The second factor relates to space discretization. If references to follow are sequential, then the obtained path is composed of small increments which cause an irregular motion. Again, taking a reference value further away from the current position, it is possible to obtain an intermediate path that is softer than the ones resulting from the sequential reference values. Figure 4 illustrates an example of this path interpolation.

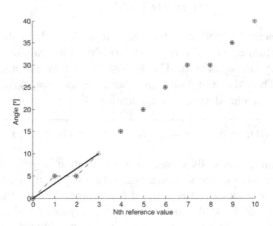

Fig. 4. Example of a path where the reference value is represented in green and the final configuration in red

Furthermore, the distance between current position and the new reference should not be significant. For higher distances, there may be loss of accuracy in the suggested path by disregarding its segments. After sending the reference value, the executed motion tends to move the manipulator joints to the defined position. However, there are no guarantees that in the next cycle this will be placed on the specified position. Therefore, at the beginning of the next cycle, the present configuration will be received from the manipulator and a new path will be calculated until the final configuration. This process is repeated until the final position is achieved. Consequently, the path is always fully executed according to the planning.

4   Results

The previous methodology was implemented and tested to compute a path between predefined initial and final configurations. The resulting path in the configuration space is shown in Figure 5.

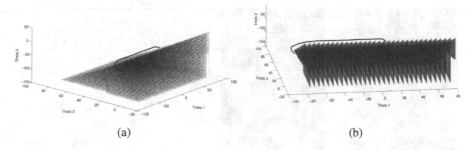

(a)   (b)

Fig. 5. Representation on configuration space of the resulting path from A* execution, considering joints 1, 2 and 3 (in degrees)

It is verified that the path obtained is fully contained in C_{free}. Therefore, there are no collisions with obstacles in work environment. This path was executed by the simulation model described in section 3.1. The results are presented in Figure 6.

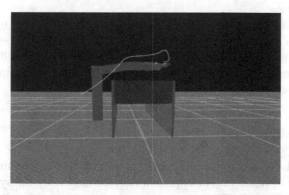

Fig. 6. Representation on the work space of the path resulting from the A* execution

Implementing this methodology is particularly demanding in terms of memory requirements. This happens because it is necessary to save information from each node related to $f(n)$ function, which, associated to the large dimension of the configuration space map, results in a higher volume of data. The path planning algorithm was applied to a set of four pairs of initial and final configurations in order to understand the execution times regarding path computing for each case. It is verified that the times obtained are formed by two components resulting from the execution of two tasks: reset of all data structures, used to save information related to the previous execution of A*, and time spent on executing the graph search algorithm. Table 1 presents the times for each configuration set.

Table 1. Execution times obtained from computing four sets of paths

Initial Configuration q_i	Final Configuration q_f	Execution Time (Reset) [ms]	Execution Time (A*) [ms]
		17,3	6,0
		15,8	5,9
		17,5	4,6
		15,9	5,4

The results presented in Table 1 demonstrate that most of the time is spent on resetting the data structure and the A* is executed in a few milliseconds. This time is compatible with the higher control level since it is lower than the feedback cycle time (40 ms) for all executed tests. Thus, the new path computation is completed before the beginning of the next cycle.

5   Conclusion

This paper developed a path planning method for cluttered environments for a robot with high degree of freedom. The approach adopted to develop the system was based on a planning of the configuration space. An approximated cell decomposition applied to the configuration space was considered. To calculate the desired path, the A* algorithm was applied on the connectivity graph. With the implementation considered, it was possible to achieve the execution times that validate the path planning algorithm. It was verified that the execution times were always lower than the cycle time of the feedback system. The decomposition cell method and the A* algorithm was applied in a high dimension configuration space. For that, some strategies used to tackle the memory constraints associated to implementation were presented.

5.1   Future Work

It is possible to identify several points that could be studied in the future. Future work can focus on the following topics:

- Implementing algorithms developed using improved processing resources and free memory available would enable improvements, such as the full representation of the configuration space and the increase of map resolution;
- Adapting the system to a dynamic work space with moving obstacles that can be used in several environments, such as situations where the cooperation between existing robots is needed;
- Inclusion of sensors that will able to perform the continuous 3D mapping of the work space, allowing the robot operation in dynamic environments;
- Implementing a trajectory planning system capable of interpolating the reference values given by the path planning system.

Acknowledgments. The work presented in this paper, being part of the Project PRODUTECH PTI (n° 13851) – New Processes and Innovative Technologies for the Production Technologies Industry, has been partly funded by the Incentive System for Technology Research and Development in Companies (SI I&DT), under the Competitive Factors Thematic Operational Programme, of the Portuguese National Strategic Reference Framework, and EU's European Regional Development Fund.

This work is also partly financed by the ERDF – European Regional Development Fund through the COMPETE Programme (operational programme for competitiveness) and by National Funds through the FCT – Fundação para a Ciência e a Tecnologia (Portuguese Foundation for Science and Technology) within project «FCOMP - 01-0124-FEDER-022701».

References

1. Latombe, J.: Robot Motion Planning: Kluwer international series in engineering and computer science: Robotics. Kluwer Academic Publishers (1990)
2. Canny, J.: The Complexity of Robot Motion Planning. ACM Doctoral Dissertation Award (1988)
3. Wu, L., Hori, Y.: Real-time collision-free path planning for robot manipulator based on octree model. In: 9th IEEE International Workshop on Advanced Motion Control, pp. 284–288 (2006)

4. Kavraki, L.E., Svestka, P., Latombe, J.C., Overmars, M.H.: Probabilistic roadmaps for path planning in high-dimensional configuration spaces. IEEE Transactions on Robotics and Automation 12(4), 566–580 (1996)
5. Fragkopoulos, C., Graeser, A.: A RRT based path planning algorithm for Rehabilitation robots. In: 2010 41st International Symposium on Robotics (ISR) and 2010 6th German Conference on Robotics (ROBOTIK), pp. 1–8 (2010)
6. Lavalle, S.M.: Rapidly-exploring random trees: A new tool for path planning (1998)
7. Kunz, T., Reiser, U., Stilman, M., Verl, A.: Real-time path planning for a robot arm in changing environments. In: 2010 IEEE/RSJ International Conference on Intelligent Robots and Systems (IROS), pp. 5906–5911 (2010)
8. Lahouar, S., Zeghloul, S., Romdhane, L.: Real-Time Path Planning for Multi-DoF Manipulators in Dynamic Environment. International Journal of Advanced Robotic Systems (2006)
9. Costa e Silva, E., Costa, F., Bicho, E., Erlhagen, W.: Nonlinear optimization for human-like movements of a high degree of freedom robotics arm-hand system. In: Murgante, B., Gervasi, O., Iglesias, A., Taniar, D., Apduhan, B.O. (eds.) ICCSA 2011, Part III. LNCS, vol. 6784, pp. 327–342. Springer, Heidelberg (2011)
10. Costa, P.G.: SimTwo (2012),
    http://paginas.fe.up.pt/~paco/wiki/index.php?n=Main.SimTwo
11. Lima, J., Gonçalves, J., Costa, P., Moreira, A.: Humanoid realistic simulator: The servomotor joint modelling. In: International Conference on Informatics in Control, Automation and Robotics, pp. 396–400 (2009)
12. Smith, R.: Open Dynamics Engine (2000), http://www.ode.org
13. Edelkamp, S., Schroedl, S.: Heuristic Search: Theory and Applications. Morgan Kaufmann, Elsevier Science (2011)

Author Index